LEPIDOPTERAN ZOOLOGY

How to Keep Moths, Butterflies, Caterpillars, and Chrysalises

Rosy Maple Moth, *Dryocampa rubicunda*
(CC Andy Reago and Chrissy McClarren)

LEPIDOPTERAN ZOOLOGY

How to Keep Moths, Butterflies, Caterpillars, and Chrysalises

Orin McMonigle

COACHWHIP PUBLICATIONS

Greenville, Ohio

Cecropia moth, *Hyalophera cecropia*, caterpillar

(CC Judy Gallagher)

Lepidopteran Zoology: How to Keep Moths, Butterflies, Caterpillars, and Chrysalises
© 2019 Orin McMonigle

Front cover image: *Euchates egle*, fifth instar caterpillars
Back cover images: (top to bottom) Pipevine swallowtail caterpillar (CC Katja Schulz; spicebush swallowtail caterpillar (CC Judy Gallagher); giant swallowtail caterpillar (CC Andy Reago and Chrissy McClarren)

CoachwhipBooks.com

ISBN 1-61646-468-2
ISBN-13 978-1-61646-468-4

CONTENTS

Atlas moth, *Attacus atlas*, caterpillar

(CC Resenter89)

ACKNOWLEDGMENTS

I would like to extend great thanks to Dr. Duke Elsner and Annie Lastar for editorial suggestions and comments. I extend thanks to countless other enthusiasts who have inspired or added to my understanding of lepidopteran zoology over the years. I am grateful to my wife, daughters Kree and Gwynevere, and son Jax for sharing my time with caterpillars. I thank God for creating these living symbols of rebirth and spectacular transformation.

IN MEMORY OF JAMES V. SMOLKA (1952-2018)

Jim earned a Bachelors in Zoology and Masters in Biology with a focus on Entomology in the late '70s. In the '80s he built the best traveling display of live and dead invertebrates (giant silkworm moths, birdwing and swallowtail butterflies, beetles, tarantulas, etc.) I have ever seen. He shared this display with the public for nearly four decades. He was a good friend who provided papers with husbandry data for caterpillar rearing and brought over his 35mm camera to help me take pictures of giant moths and caterpillars prior to the age of the digital camera.

Oakworms, *Datana* sp., are attractive display
caterpillars that turn into tiny, brown moths

LEPIDOPTERA IN CULTURE

The butterfly net has long been the symbol of the entomologist. Of all the insects, this group of creatures has spurred the excitement and imagination of the collector and nature lover like no other. *Lepidopteran Zoology* is my contribution to the appreciation and study of this variable group of some 180,000 described species.

My earliest memories of invertebrate zoology are tied to the Lepidoptera. One memory involved storage of Luna and Polyphemus moth cocoons in an old cigar box. A few months earlier I had collected eggs from wild females and reared the caterpillars in small terrariums. At the time, entomology enthusiasts who did not have the resources for glass-covered insect drawers were instructed to hunt down wooden cigar boxes to store dried insect speci-mens. The cigar box also seemed like a great place to store giant silk moth cocoons. Predictably, just before the onset of winter I discovered peculiar giant silk moths running around my room. I had never seen moths like these before. They resembled giant silk moth species I was familiar with, but they were pale gray in color and much smaller. An investigation of the cigar box revealed what species they were and where they had come from. This taught me a few lessons. First, an overwintering strategy is a must for any lepidopterist. Second, I cannot always expect the creatures I rear to be as big or colorful as the wild specimens. Some foodplants can be used which allow caterpillars to survive and grow, but still do not allow for optimal growth or a high rate of survival.

Another fond, early memory was bringing cecropia moth eggs to Florida during a month-long trip and returning with grown caterpillars. I had saved up money from my paper route for a plane ticket so I could stay at my uncle's house for a month. My singular goal was to observe and hunt invertebrates unlike anything I could find twenty miles south of the Canadian border. Before leaving I collected eggs from a cecropia female and knew they were going to hatch long before I returned. I was certain the tiny larvae would starve to death without care. If I glued the eggs to a leaf outside, the prognosis would almost certainly be the same and I did not have a large indoor screen cage. When they hatched, I realized the native pines and palmetto in the Orlando area were useless as food, but I was lucky to find a single tiny maple tree planted in a front lawn a few houses away to feed the caterpillars. This sustained them through three or four molts and the return trip. Of course I would never repeat such a thing in today's world, but it was long before September 11 and I was 13.

The mystery of the painted caterpillar leaves taught me an early lesson about humidity and ventilation and the adaptability of caterpillars. I had a plot of foodplants growing in a glass-covered terrarium (tiny citrus "trees" from orange and grapefruit seeds) to feed giant swallowtail butterfly caterpillars.

The swallowtail caterpillars ate most of the leaves off the citrus seedlings during their first three instars. In order to provide an adequate food source, the fat caterpillars were moved to a relatively large citrus houseplant sitting on my dresser. Later that day, I looked at the citrus tree and did not see a single caterpillar. I searched around the room and under the furniture with no luck. Some time later I searched the citrus again. I was surprised to see what appeared to be two-dimensional caterpillars painted on some of the leaves. Caterpillars are like vascular plants in that they can adapt their external surface and respiration to high humidity and limited airflow. Like plants, they rapidly lose moisture if moved from low airflow and 100% humidity to open ventilation and the very low humidity of a home. However, a caterpillar does not have near the structural integrity of a plant.

There are countless resources for people who keep Lepidoptera and zoological literature has been available to the public for more than four hundred years. *The Silkewormes and their Flies* (Mouffett 1599) discusses experiences in Italian silkworm culture and the infectious nature of silkworm diseases. Merian (1679) published drawings depicting the life cycles of fifty different moths and butterflies, followed by Vol. 2 in 1683 and *Metamorphosis insectorum Surinamensium* in 1705. Her work was inspired by the silkworm, whose life cycle is the first plate in Vol. 1. Warrington (1841) discussed lepidopterous larvae that do or do not eat their eggshells, followed by various articles over many decades in *The Entomologist*. *Caterpillars and their Moths* (Eliot & Soule 1921) is a three-hundred-page text on moth keeping. A *Silkmoth Rearer's Handbook* (Cooper 1942) is an impressive text focusing on saturniid rearing, later updated in 1956 and 1982 by different editors and various contributors from the Amateur Entomologist's Society (United Kingdom). *Moths and How to Keep Them* (Villiard 1969) contains amazing details on diet and development of a huge variety of North American and exotic moths. The number, variety, and quality of books published in the last fifty years is legion, including massive compendiums devoted strictly to moths, butterflies, or caterpillars alone.

The first book I remember pouring over was an English adaptation of *Kaiko* (Johnson 1982 from Kishida 1977). This small book is illustrated with excellent developmental photographs of silkworms which piqued my interest. From the age of 11, every morning while delivering papers I checked the lights around the apartment buildings to find giant moths (some were very high up and required the acquisition of very long sticks). As I progressed in my husbandry endeavors, I used field guides to determine possible foodplants. However, I ran across nothing that offered husbandry information at the local libraries. In the early 1990s I acquired copies of a few pages of Villiard's fantastic moth text from Jim Smolka, and finally acquired a used copy of the book in 2010. The content and quality of husbandry data seemed like something from the future rather than the past.

The intent of *Lepidopteran Zoology* is to introduce the beginner to zoological displays, complexities of maintaining Lepidoptera in captivity, and to inspire greater interest in these fantastic creatures. The species section is compiled of updates to husbandry articles published in *Invertebrates-Magazine* over the last two decades. The stories and husbandry experiences will hopefully prove informative and inspiring to the beginning lepidopterist. They may provide some interesting data to the seasoned lepidopterist as well.

Moth and Butterfly Houses

Public Lepidoptera exhibits probably first began in England with displays of large saturniid moths, documented at least as far back as 1881 at the Zoological Society's insect house (Cribb 1982). Modern exhibits seem to be focused almost exclusively on butterflies. There are currently three butterfly conservatories in the United Kingdom, while other parts of the Commonwealth like Australia and Canada have as many or more. The world capital of butterfly houses is the United States, with more dedicated facilities than all the countries of Europe and Asia combined. Butterfly World, which opened in Florida (USA) in 1988, is the world's largest butterfly park. It is difficult to say the exact number of butterfly houses in the United States today, but there are at least a hundred. Still, an internet search for butterfly houses does not pull up most of the ones I have visited. The oldest continuously-running, indoor butterfly exhibit is the butterfly room at Insect World in Cincinnati that has been in operation since 1978. Cleveland Botanical Garden has displayed butterflies, including the iridescent blue morphos and gigantic owls, year-round in its 'Costa Rican Glass House' since 2003. Butterfly Pavilion, an indoor greenhouse-tunnel of butterflies at the Smithsonian in Washington D. C., opened in 2008. Seasonal exhibits are probably more common than year-round displays. The butterfly museum on Put-in-Bay (an island in Lake Erie) is among the many that follow the summer tourist rush. Many zoos in the north stock greenhouses full of butterflies only during the summer months.

At its core each butterfly exhibit has a chrysalis house. This small structure has a roof, walls, and a floor. It often has multiple floors and looks something like a dollhouse. The front can be fully open or covered in screen or

White peacock, *Anartia jatrophae*, in a butterfly house

Checkered white, *Pontia protodice*, caterpillar in a butterfly house

Common buckeye, *Junonia coenia*, in a butterfly house

Julia heliconian, *Dryas iulia*, in a butterfly house

glass. The floors have cork bottoms or multiple openings, each used to hang a chrysalis or cocoon. Spacing is required to provide enough room beneath and on each side for the wings of the adults to expand. The floors are usually angled and have a gap so the animals can move down to a large exit in the back or side when they are ready to fly. There is usually an exit so specimens can leave the cramped chrysalis house to venture into the surrounding greenhouse. However, the emergence area may be blocked where chrysalises or cocoons are wild-caught, in order to contain their putative parasitoids. In such cases the exits are screened, and parasitoids (usually wasps, seldom flies) can be directed towards the lighted neck of the bottle for easy removal. Sealed chrysalis houses must be checked every day and the inhabitants released so they do not beat their wings to pieces in the cramped space and damage other emerging specimens.

Nearly every butterfly house purchases chrysalises from vendors rather than rearing their own. This is partly because the time and space involved in continuous rearing greatly exceeds that of an exhibit and partly because an inordinate level of devotion and unique ability are required to continuously rear most species. Any devoted keeper can maintain a species through part or all of a life cycle, but few are capable of continuous culturing. Even with great skill and devotion some species still require intermittent infusion of new stock, which can be difficult outside of the natural range. In the case of exotics like morphos the butterfly farms tend to be in the natural range of the species produced. Even the most fantastic individual specimen is relatively inexpensive, but most live less than a month and a large number of specimens must be purchased on a continuous basis to stock an exhibit.

Chrysalis house

Locked chrysalis station in a butterfly house

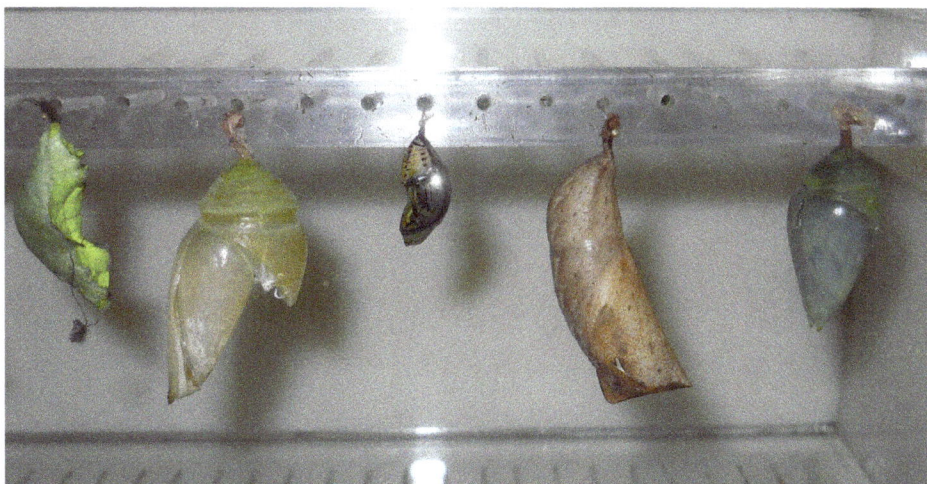

Chrysalis attachments in chrysalis station

All lepidopteran exhibits are designed to keep their residents contained. The extent of security is partly determined by whether the inhabitants are native or foreign. Lessened effort is usually placed on preventing escape of native species, but efforts are worthwhile to prevent loss of costly exhibit animals. When it comes to exotic species the import permits usually include regulations which require multiple barriers to prevent escape. There is almost always a walkthrough curtain of plastic chains to knock butterflies off exiting visitors and a second set of doors to create a vestibule for containment. In addition there may also be a downward and backward air curtain that automatically turns on when the inner door is opened. Doors for both the entrance and exit have signs posted which direct the visitor to check themselves in a mirror for accidental hitchhikers before exiting the next set of doors. If the butterfly house displays exotic species, plant materials and soil removed from the facility are usually autoclaved to ensure eggs or hitchhiking organisms (if present) are destroyed.

Loss does not always equal escape. During its opening years (1992-1993) condensation on the glass at the Cleveland Rainforest resulted in nearly complete loss of the exhibit animals. The insects were stuck to the glass like giant flies on oversized sticky paper. Netting was installed to cover the glass, but many butterflies still found their way behind the netting. The idea of butterflies flittering around the "rainforest" was abandoned and moved to a seasonal exhibit in a small greenhouse that still runs today. Visitors also pose some risk to the butterflies, most often children that grab or step on them. In 2016 a woman was arrested for stealing an owl butterfly from the Cleveland Botanical Costa Rican exhibit.

Owl butterflies at feeding station in a butterfly house

For me, the best butterfly house should have caterpillars, but they almost never do. In the opening year in 1995, Atlas moth caterpillars could be seen eating a few types of plants in the greenhouse at the Cleveland Zoo's seasonal exhibit. The gigantic moths were fantastic to see in a butterfly house. They were hugely impressive even though they did not fly around during the day. Seeing their gigantic, beautiful caterpillars munching away was even more fantastic. Unfortunately, the Atlas moths were never purchased again since the greenhouse plants were far more important to the curators. I do not remember seeing caterpillars in a contained butterfly exhibit again until I visited the butterfly greenhouses at Epcot in Florida in 2018. The large number of caterpillars representative of a few species seemed to be purposeful and impressive. This does not mean caterpillars are rarely featured in exhibits, just not the flight areas of butterfly houses. For many years there has been a *Manduca sexta* exhibit at the Smithsonian insect zoo in a large glass tank that features the massive hornworms and giant pupae. The huge, colorful caterpillars of butterflies and giant silk moths are commonly featured, temporary displays at nature centers.

Many botanical gardens, zoos, and parks have areas with butterfly garden signs, but they do not stock or rear any specimens. Nature provides. Of course these "butterfly gardens" are almost never even remotely so impressive as a poorly stocked butterfly house, but they can be very enjoyable and busy with all manner of pollinating insects.

I have walked through many butterfly gardens on perfect sunny days without seeing a single lepidopteran, though they might be rather lively at the right time. Butterfly enthusiasts, parks, and botanical gardens often plant

Automeris io caterpillars on display at a nature center

Temperate, day-flying moth in a butterfly garden

White in a butterfly garden

flowers to attract butterflies. There are many resources (books, articles, and webpages) that list plants that can be grown to attract butterflies, but almost any natural flower will attract some type of lepidopteran. Flower hybrids like doubles and roses with an unnatural numbers of petals may not allow for nectar feeding. Butterfly gardening does make it possible to enjoy the beauty of butterflies without all the work and without facing the trials of captive rearing or caterpillar farming. Yet it lacks the joy and sense of achievement found in successful culturing. Each stage and species has its own unique quirks, even if is only the changing appearance between caterpillar instars.

Lepidoptera Culturing

Acquiring livestock from vendors is a possibility in some countries, but it is difficult or requires impossible-to-acquire permits in others. Vendors in many European countries have seasonal availability of Atlas cocoons and countless other fantastic exotics. In the U.S., living silkworms, hornworms, and waxworms are readily available as food for reptiles. Certain caterpillars, chrysalises, and kits are available from a few butterfly nurseries, but only a small number of species can be purchased, and they can only be shipped to certain states. Fortunately, the Lepidoptera is an order of insects where a person living nearly anywhere on earth can find an impressive range of spectacular and gigantic species. These can provide new and interesting entomological experiences for a lifetime. A small number of species are protected (essentially very rare butterflies like blues and birdwings) so rearing these would require permits even if a lepidopterist could acquire specimens from their own back yard.

Husbandry requirements have similar and consistent themes, so an experienced lepidopterist can usually be successful adjusting to differences when rearing unfamiliar species. Still, even the most successful keeper will encounter species they find difficult, even some that other keepers find easy. Maintaining lepidoptera requires a natural ability to keep things alive as much as it requires useful rearing guidelines, acceptable environmental conditions, and the best food plants. The experiences and information listed in this text can be used by the beginner to greatly improve chances for success and to understand where they may have gone wrong.

Collecting and storing eggs ranges from very simple species to trying and difficult ones. Many commonly reared moths can be placed in a paper bag in which they will eventually lay all their eggs. The bag is then cut apart and eggs clusters are placed on food plant cuttings. Some eggs will have to be moved to hatching or storage containers. Most butterflies and certain moths like sphinx moths and leopard moths usually will lay few to no eggs in an empty paper bag. They prefer to deposit ova on fresh leaves of the food-plant or a nearby surface. These species will need to feed and fly around in a large enclosure since they may only lay a few eggs each day and may only lay eggs on specific host plants. For many butterflies it is important to keep the leaves fresh because the eggs will desiccate almost immediately when the leaves wither. Spraying the eggs with water usually drowns and kills tiny ova. If the leaves have wilted, cut portions with eggs can be laid on fresh leaves, but the hatch rate is likely to be very low. It is important to time cuttings so they remain fresh or use potted plants for the first instars. Most eggs hatch in a few days to a few weeks, but some, like silkworms, can be refrigerated for hatching the following spring. These can be stored in a covered container in the vegetable crisper drawer and occasionally misted to prevent desiccation (the eggs of large moths usually survive misting with water).

The eggs of wild-collected, adult females are often fertilized, but fertilization can be a huge challenge for moths and butterflies that emerge in artificial enclosures. Adults that eclose indoors often refuse to mate because the stimulus for courtship is missing. Greenhouse-sized flight cages can solve the problem for many species, but temperature, lighting, food, or other impetus may still be lacking. The pheromones produced by female moths are partly determined by chemicals extracted from the foodplants (Arnett 1993), so substitute foods may reduce or eliminate the male's incentive. Hand pairing is common practice for giant silkmoths and swallowtail butterflies (Watts 1996, McMonigle 2007), but they can die from the stress of handling, or lay infertile eggs despite intensive effort. Infertile ova look the same as viable ova when first deposited, but commonly collapse after a day or two. The collapse tends to be severe, so it is easy to tell from the slight dimple found in the center that some eggs display during normal development. There are lepidopterans, famously gypsy moths and corn root borers, that do not require males or need to mate. They can reproduce through parthenogenesis. However, most species kept by lepidopterists require a male counterpart.

Hand pairing involves holding a female in one hand and a male in the other. The wings are held together to prevent beating and the legs may hold onto a finger or hang free (some males are more or less distracted when grasping something). Males have a large clasper at the end of the abdomen and are often thinner near the middle. A male's clasper is usually easy to see, but is obscured by thick scales. The male's clasper is separated with a fingernail or a thin, upright pin. The pin (or metal wire) is inserted in wood or Styrofoam to keep it upright. After manually opening the claspers a few times, they should stay open, until the end of the female's abdomen can be placed in-between. If successful, the abdomens will stay linked together and the animals can be gently placed on a stick or piece of wood. If they separate in less than an hour, hand pairing should be attempted again. Certain species and specific individuals are more likely to accept this treatment. Also, the age of the moth or butterfly is important. Hand mating is usually only successful for saturniids within a few days of emergence. Old specimens may mate successfully but die after handling or lay few to no eggs.

First instar larvae hatch a few days or weeks following oviposition, or after being brought to room temperature following cold diapause. Eggs laid at the same time usually hatch within hours of each other and many caterpillars wander off to feed together, at least at first. Hatchling caterpillars are tiny and have a difficult time finding food unless they are placed directly on the correct leaves. Ventilation ports should be covered in fine screening such as muslin cloth. Even the largest hatchlings can be small enough to crawl through standard window screening. Hatchlings often wander for the better part of a day before starting to feed, but this depends on how acceptable the foodplant is.

Adequate ventilation and humidity can be difficult to provide and varies across species and stages of the same species. Tiny caterpillars, less than 1/2" (1.3 cm) long, usually do well with limited ventilation and relatively high humidity (standing water should be avoided, as even a thin film is deadly).

Callosamia promethea, 2nd instar caterpillars

The larvae of miniature Lepidoptera may never grow large enough to require increased ventilation. If early instars of large species are kept with limited ventilation, increased airflow and decreased humidity should be stepped down slowly by the third or fourth instar. Caging for large caterpillars should have low humidity along with good ventilation and airflow. A fan blowing across the vents is helpful for very large larvae if ambient humidity is high.

The required ventilation increases with size even if a molt has not taken place. A freshly molted caterpillar is a fraction of the size it will become. The frass grows proportionately larger between and within instars. Waste from an early 4th instar caterpillar is significantly smaller than frass produced by the same animal in late 4th instar. If frass appears wet or becomes moldy, within a short time the caterpillars will turn into limp, black bags of fluid. Disease and death are common with excessive moisture. Certain food plants result in watery frass, so changing to a food plant that results in drier frass can eliminate or reduce problems. If adequate ventilation is difficult to provide, or the ambient humidity does not allow frass to dry out despite ventilation, the frass will need to be removed and discarded a few times a day. Excessively dry conditions can also be harmful and result in death or molting difficulties. This can be solved by covering a portion of the ventilation. Misting dehydrated specimens with water may help a little at first, but regular misting results in extremely high or total die-off.

Every container imaginable has probably been used to house Lepidoptera. Tiny specimens are often contained in petri dishes or small plastic cups with a few pinholes for ventilation. For large caterpillars, full screen cages or open trays work well. Pop-up screen cages are commonly used to house butterfly

Stems in containers keep leaves fresh

larvae. An economical option for batches of extremely large caterpillars (big Saturniidae) is a wide, cardboard box. Packing tape is placed around the upper rim and the smooth surface is coated in petroleum jelly to prevent escape. Shallow trays are used to catch falling frass, but do not prevent escape. The bottom inch or two of the foodplant stem is coated in petroleum jelly to keep the caterpillars from crawling down the water container and across the tray to freedom (and almost certain death). Still, some specimens can get past the slippery stem so a second level of containment is helpful. There are 6" (1.83 meters) screen, pop-up tents everything can be placed inside to offer a third barrier.

Caterpillars molt in a way that reduces the effects of problematic molting. Instead of the old head capsule splitting open during the molt, the new one develops behind the old head capsule. The old capsule is slowly squeezed forward and falls off in one piece. Many beetle larvae have similar looking head capsules, but they split between the lobes and frons. Beetle larvae die from starvation if the new head becomes stuck, because there is no way to recover from mouthpart deformities. If caterpillars are kept too dry the head still molts correctly, but the rest of the body becomes stuck in the old exuvia. This is one of the only cases in invertebrate husbandry where a keeper can actually help an animal that has molted badly. The old skin can be carefully lubricated with water and pried or very carefully cut apart and removed. Subsequent molts should proceed without difficulty if conditions are improved. Changing the type of foodplant, keeping leaves fresh, or increasing ambient humidity can help prevent the problem.

It seems like it would be a simple task to cultivate native caterpillars outside by simply placing them on their host foodplant and collecting the

Waved sphinx, *Ceratomia undulosa*, caterpillars getting ready to molt

full-grown animals before they wander off to pupate. However, the survival rate outdoors without protection is commonly close to zero. Birds, spiders, and other predators are very efficient. Parasites and diseases also decimate wild populations. The average lepidopteran female produces hundreds of eggs, but only two offspring (on average) will survive to maintain a consistent wild population. Sleeving is a popular way to farm caterpillars outside that provides protection from predators. The host plants leaves are beaten to remove predatory invertebrates, a bag of netting is pulled over, caterpillars are placed inside, and the end is tied. Muslin cloth is commonly used because the netting is fine enough to keep out parasitoids. Sleeving can still be labor intensive since the first two instars are usually still reared in containers. Also, large caterpillars have to be moved periodically to new branches before the leaves are mostly consumed. In my yard I have had poor luck sleeving because *Zelus luridus* assassin bugs will wait for caterpillars to get close to the screen. The assassins drain the caterpillars of bodily fluids through the netting. I have also had an animal tear open the netting during the night.

Wandering behavior is a common problem that can lead to starvation or early pupation. Specimens that pupate early are more likely to die as a chrysalis or become very small adults. If the food begins to dry out, leaves become mostly consumed, or the type of food is switched (even new branches of the same food at times), caterpillars often climb the sides of the enclosure or run around on the floor of the cage. They are surprisingly unable to find fresh leaves even if the optimal food is almost touching them. If they are not on touching stems, the food may never be found. Caterpillars usually move over

This wild *Manduca* caterpillar developed wasp parasitoid cocoons.

to fresh branches on their own if the fresh leaves are placed touching the old stems or branches. If caterpillars are found on the bottom they should be manually placed on the fresh leaves. Care should be taken not to damage the fleshy pads of the prolegs or handle too often. If a caterpillar is about to molt it will not move from the old leaves and it should be left alone until it has completed the molt. A pre-molt caterpillar can usually be identified by its posture (straight and slightly lifted at the front), lack of movement, and the presence of a fine silk molting pad on the surface it is resting on. Also, the head shape of a molting caterpillar looks unusual as the new head capsule is developing behind it.

The chosen food plant can make a big difference on survival and growth. While most species accept a narrow range of leaves from different plants, one usually works better than another. Less desirable food will be eaten with hesitance. Preference seems to be generational—when I switched sphinx moths to hybrid privet the first generation ate hesitantly but the following generations did not hesitate. One might think that feeding multiple types of food would provide a wider array of nutrients and theoretically better growth, however switching between acceptable foods usually results in reduced feeding and reduced adult size.

The number of caterpillars to be reared should be considered ahead of time. If too many are reared at once there may not be enough foodplant available. Problems with overcrowding and disease are greatly increased if hundreds or thousands are reared at once. Too many caterpillars can exhaust and drain a keeper enough to force him or her to give up, or reduce care to the point where everything dies anyway.

Feeding the adults is usually much more difficult in captivity than feeding larvae, even though the quantity of food is significantly less. Many moths, including the giant silk moths, do not feed or drink, but sphinx moths, other moths, and most butterflies will die without laying eggs if they cannot feed. If a greenhouse with the right kind of flowers is available, feeding may be easy, but most lepidopterists do not have such extravagant resources. Some species will readily accept nectar cups, insect jellies, or the juices from freshly sliced fruit.

The overwintering strategy can take time, research, and testing to perfect for a given species. A number of butterflies overwinter as adults (famously the monarch and ladies), but will not live very long indoors or at room temperature. Many temperate butterflies overwinter as chrysalises, the same as temperate giant silk moths. Chrysalises can be stored through the winter months in a cold box outside, or within a container in the refrigerator. It is important to avoid cooling before the caterpillars transform to pupae, but they should be cooled before they begin to develop for eclosion. It is too late if the wing pads of the pupa turn light or dark. The pupae of some species like *Eacles oslari* and *Acronicta americana* will stay in diapause at room temperature and emerge the following summer. Silkworms, most copper butterflies (*Lycaena* spp.), and underwing and owlet moths (Superfamily Noctuoidea) are among those that overwinter as eggs. Silkworm eggs can only be kept cold if the day cycle the female moth was raised under was short (12 hours or less of daylight is best to induce diapause). Cooling eggs that are going to hatch will kill them, or the eggs will hatch in the refrigerator. Tiger moths (Arctiidae) usually overwinter as full-grown caterpillars, while skippers (Hesperiidae)

Junonia coenia feeding

can be full- or half-grown larvae depending on the species. Lappet moths (Lasiocampidae) can be eggs, caterpillars, or pupae depending on the species. Tropical species may not have a cold diapause, but some like *Attacus* pupae will delay eclosion for half a year or more if kept very dry. Of course they should not be kept extremely dry or they lose too much internal moisture and cannot eclose.

Proper emergence of adults is usually related to the humidity the chrysalises or pupae are kept at. Most butterflies emerge from their chrysalises in a short time so ambient humidity is not significant, but they can be misted a few times a day in very dry conditions caused by central air or heat. Temperate moth pupae often must be cared for over a period of six to nine months. Too little humidity for too long will result in death or emergence of specimens that cannot expand their wings. Most cocoons or naked pupae can be kept in damp paper towel inside a plastic container with limited airflow (a few pinholes). Dirt or leaves can develop mold and result in suffocation. Species that incorporate leaves in the cocoons should be carefully cleaned of leaves.

Species emerging from underground face a difficult physical challenge to reach the surface. Sphingids sometimes fail to expand their wings if they are allowed to sit as a naked, unearthed pupa for emergence. Soil-pupating species are not easy to keep buried their entire pupal period underground because it is difficult to provide the correct environmental conditions—too dry and they desiccate, too wet they die from fungal infections. The larvae can be provided enough soil to dig deep and then can be unearthed after there has been sufficient time to complete pupation, or they can be provided 1/2″ (12.7 mm) of soil so the pupa is not formed underground. Pupae are

Junonia coenia starting to spread its wings

Manduca pupa

then stored in a refrigerator or kept at room temperature depending on the plans for dealing with the adults. They can be stored in plastic containers, with each pupa rolled inside a piece of soft paper towel. When it is time for adult emergence, the pupa remains in the rolled paper towel. This can be kept relatively tightly rolled by securing the loose edge with tape. This gives the emerging adult just a bit of resistance on the way out of the roll. Remarkably, just that little bit of extra effort seems to be enough to assure a high percentage of them fully expand their wings (pers. comm. Elsner, D., Michigan State University Extension).

Handling

Many lepidopteran eggs can be safely handled without fear of damage, especially the sturdy ova of large moths. The tiny eggs of small butterflies and moths can be smashed during handling, but damage is usually only a concern if ova are removed from the surface they were laid on. The leaf or surface should be removed rather than the egg itself. Most eggs are glued to surfaces with water soluble secretions so they can be moistened if removal is required. An artist's paint brush can be used to transfer delicate eggs with moderate success.

Caterpillars at all stages can be handled safely but many lepidopterists avoid handling larvae whenever possible. Touching specimens with hands can lead to death and disease even if the creatures seem to be unhurt. Poor handling can damage the prolegs which leads to excessive bleeding, rapid death, or entry of disease-causing organisms. Very tiny specimens can easily be torn apart or smashed. Even when specimens were handled patiently and carefully,

Antheraea polyphemus, 5th instar caterpillar

the disturbance can still lead to wandering behavior and starvation. Usually, feeding specimens will move to fresh food on their own unless they are about to molt. Molting specimens should not be touched. If the caterpillar is stationary, the shriveled leaves or stems can be cut around the animal with scissors. The small pieces of leaf or stem can be laid on a fresh leaf. If a molting animal has lost grip of its silk molting pad (the molting pad is very difficult to see even for larger caterpillars) it can be placed on an old cocoon or similar surface that allows for an easy grip. The prolegs must be attached to a surface for a successful molt.

As mentioned, caterpillars are one of the only arthropods that can be successfully helped out of a stuck molt. If kept too dry, the head can still molt correctly, but the rest of the exoskeleton can become stuck. The old skin is then sprayed with water for lubrication and carefully torn or cut and removed. Trachea linings look like white string or floss connecting the old and new spiracles. Care should be taken to remove as much of the trachea lining as possible by pulling very slowly. Not all of the lining will make it out of the spiracles. Trachea lining that breaks off inside leaves the caterpillar more sensitive to excessive moisture and limited ventilation. It is important to be especially patient when removing the exoskeleton from around the prolegs. The prolegs should be given time to release grip on their own. Tears in the prolegs lead to excessive bleeding that can stunt growth or kill the caterpillar.

Handling the chrysalis or pupa is usually safe unless it is accidentally dropped on a hard surface. Moth cocoons are often cut open to determine genders and check on the health of the pupae. Opening the cocoon does not harm the pupa inside unless the pupa is smashed or cut in the process. If the

Papilio polyxenes

ventral surface of the fourth segment of the abdomen back from the wings is solid it is a female; if there is a large indent or break in continuity of the surface, the specimen is a male.

Handling adults carefully is unlikely to kill them, but they lose scales with reckless abandon. The beautiful colors and patterns of scales on the body and wings are easily marred with even the most gentle handling. Captive-reared giant silk moths and swallowtail butterflies are often hand mated (held abdomen to abdomen until pairing is achieved) which often severely damages their appearance. Hand pairing is a last resort since (even if the animals survive and pairing appears successful) the eggs can still be infertile.

ARTIFICIAL DIETS AND ADDITIVES

Many lepidopterans have been successfully reared on artificial diets. Continuously cultured species like painted lady butterflies, tobacco hornworms, and silkworms are commonly reared on specifically formulated artificial diets. Gypsy moth diet is also available and can be used for tobacco hornworms. The main ingredients in artificial diets include agar as a thickening agent, casein protein isolate, sugar and wheat germ for primary nutrients, methylparaben as a mold inhibitor, and powdered leaves of the foodplant to provide attractant/flavor. Cellulose powder is a common, but indigestible, ingredient. Various vitamins, acids, and oils can be added but there has been very little testing to determine if they help with improved survival or growth. Waxworm moths (*Galleria mellonella*) are usually reared on a simple diet of sugar, honey, and wheat, but more complex mixes may work better. Homemade recipes for most species can require a bit of testing and tweaking to work at all. Even with a good dose of methylparaben, homemade mixes seem to mold more readily than commercial diets. Unfortunately, commercial diets are available for few species and are relatively expensive. Finished media is even more expensive than dry mixes due to shipping costs. Ready-to-use media has a short shelf-life compared to dry mixes. Dry ingredients of artificial diets can be mixed and stored safely for months or years. The dry ingredients are mixed with water, boiled to activate the agar, poured into the rearing container, and left to cool and set up for at least an hour. The completed media should be set up just before eggs are about to hatch because it fouls or dries out after a few weeks.

The only artificial diet that is very inexpensive and also easy to use is sliced potato. Unfortunately, it is only good for a few species of hawkmoth in the last instar (specifically *Manduca* that feed on tomatoes and potatoes), but it is amazingly simple. Hornworms surprisingly do not try to wander from the food and the frass takes a long time to become moldy.

Additives can improve growth and survival, but mostly are useful to prolong the viability of leaves that wilt quickly or are refrigerated. A simple mix is a tablespoon of powdered milk (for casein protein) and a pinch of baker's yeast dissolved in a half cup of water. The liquid is added to a small spray bottle and applied to the leaves to be used. Stems are cut, placed in water, and the leaf's surface is allowed to dry before offered to the caterpillars. A mixture of milk and yeast must be used in 24 hours or it really starts to stink, but the ingredients are easy to find. A mixture that may work similarly, or better, and should have a longer life is 2 tablespoons of dextrose, 1.5 teaspoons potassium nitrate, 2 tablets of brewer's yeast, and 1 pint of water (Villiard 1969).

Manduca caterpillar feeding on a potato slice

Hylaphora cecropia

MOTHS

CECROPIA MOTH
Hyalophora cecropia (Linnaeus, 1758)
Also known as the robin moth due to its monstrous size, *H. cecropia* boasts the largest wing area of the North American giant silkmoths (family Saturniidae). *Hyalophora euryalis* is a similar species found on the west coast that has a darker reddish background color and an elongate discal spot on the hind wings. Female cecropia are the larger of the two sexes, but even male moths can exceed six inches (15 cm) in wingspan. Caterpillars are likewise gigantic and can grow to five inches (13 cm) in length. Though larvae are massive, they are not the largest North American caterpillars (those of the regal moth, *Citheronia regalis*, likely hold that distinction).

The night-flying female has only about five days to mate and lay 200-400 eggs. In nature a male often finds the female the night she emerges. Eggs hatch in ten to twelve days and caterpillars pass through five instars in about six weeks. 1st instar larvae are black, 2nd instar are yellow with black tubercles, while the last three instars are green with yellow, blue, and red tubercles.

Hylaphora cecropia caterpillar

Hylaphora cecropia
building cocoon

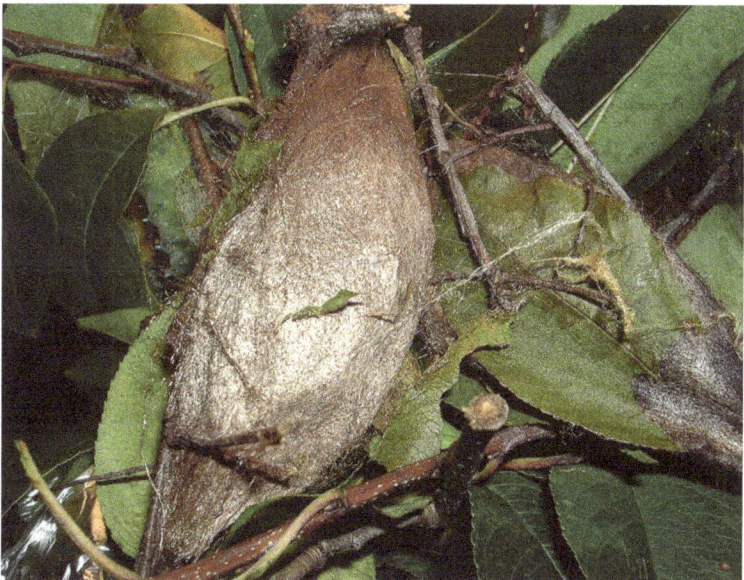

Hylaphora cecropia
cocoons

Once it has finished eating, the 5th instar takes a number of days to build a silk cocoon. Afterwards, another week is needed for the caterpillar to molt to the pupal stage within the cocoon.

Cocoons come in a number of forms. The first is egg-shaped and surrounded closely by an outer sheath and extensive silk anchoring that obscures the shape. In the second form, the outer sheath is expanded into a large, globular, bag-like secondary cocoon that loosely encases the inner cocoon and silk lines. Wild-collected cocoons of the second form are often larger than a man's fist. The cocoon forms seem random and do not correlate with gender, population, or foodplant.

Although only certain plant leaves can be used as food, there are more choices for rearing cecropia caterpillars than for other species. I have had good results with privet, lilac, maple, and wild cherry. Other common food sources include apple, ash, birch, plum, and willow. If the type of food is changed, they may eat very little for the first day or two and some may not recover. Branches should be cut and placed in water to keep the leaves fresh. Leaves should be replaced when they begin to dry out or the caterpillars may stop feeding.

It is difficult to rear large individuals indoors. Overcrowding and withering foodplants signal mature larvae to spin cocoons prematurely. However, the indoor survival rate can be better than 90% with good husbandry. Enclosures should be well vented and should never be misted. Outdoor rearing can result in much larger moths. They should be moved at 3rd instar to sleeves for protection from predators. If caterpillars are placed on plants, without protection, the survival rate is usually very close to zero. Considering the

Hylaphora cecropia differently shaped cocoons

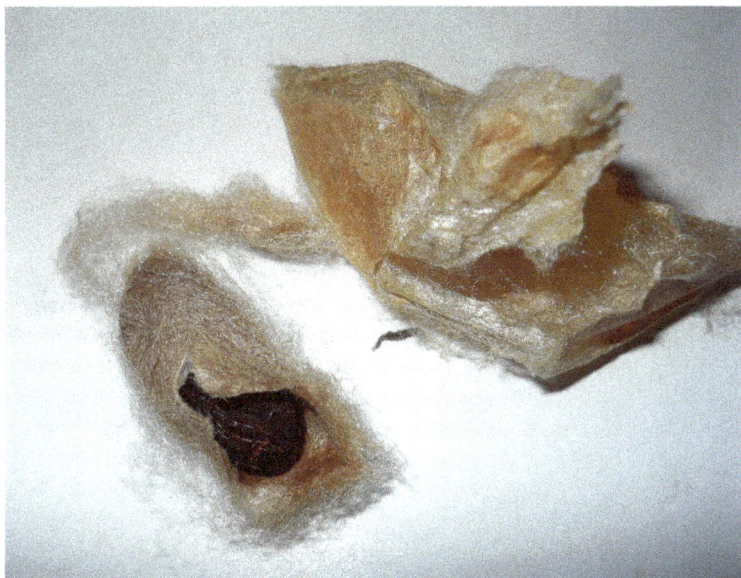

Hylaphora cecropia inner cocoon removed and opened to show pupa

Callosamia promethea male approaching female

number of eggs produced by a female, wild survival rate is only half a percent, assuming the population is relatively stable.

Promethea Silkmoth
Callosamia promethea (Drury, 1773)
Callosamia promethea is a member of the giant silk moths (family Saturniidae). It belongs to the tribe Attacini that includes *Attacus, Hyalophora, Samia,* and other genera. Two similar *Callosamia* species are found in the U.S., *C. angulifera* and *C. securifera*. Promethea is a medium-size species growing to four inches (10 cm) in wingspan. Cocoons and caterpillars are small for the size of the moth, so large groups can be reared with far less effort than most saturniids. Promethea's relation to the Asian Atlas (*Attacus atlas*) and cynthia (*Samia cynthia*) moths is noticeable as they likewise enjoy privet and have a relatively large wing area in relation to the size of the cocoon and caterpillar. Older caterpillars of all three have a white, powdery coating. The cocoon structure is also similar.

Anyone who has found a giant silk moth at night and released it the next day learns rapidly why they fly after the sun has gone down. Almost immediately upon release a bird swoops in to eat it. Male promethea are one of the few moths that fly when the sun is still up. The smaller, dark males are believed to resemble the coloration of pipevine swallowtail butterflies, which keeps them alive as they flutter across sunlit fields late in the day. The bulkier females are reddish brown to pink, and fly when it is dark.

Though a common species, I did not encounter wild promethea until 2004. My wife pointed out a female on the wall of our front porch. It is

Callosamia promethea female laying eggs

always appreciated when someone finds me a nice moth, especially when it is something I have never seen before. Over the years I have run across hundreds of giant silk moths, primarily Luna, cecropia, Io, and Polyphemus, as well as regal and imperial, but this was my first promethea.

I placed the female in a paper lunch bag for egg laying. After two days, she did not lay any eggs. This seemed very odd since female saturniids usually start laying eggs about the time the bag is closed. On the third day I placed a small piece of golden privet in the bag and she began laying eggs after a few hours. 160 were laid over the next three days.

The eggs hatched into nearly black 1st instar caterpillars. Plum, lilac, and privet were tried and the privet was readily eaten. 2nd instar caterpillars are ringed in black and white and are very reminiscent of monarch butterfly caterpillars. 3rd instar caterpillars are mostly white with yellow thoracic tubercles, but some retain black banding. 4th and 5th instar caterpillars are green and white with red tubercles and looked like plastic toys. At maturity, the fifth instar caterpillars stopped eating and spun cocoons woven into a spindle shape, similar to cynthia cocoons.

STAGES IN DAYS AT 75°F (24°C)

Adult	*7*
Egg	*12*
1st instar	*6*
2nd instar	*8*
3rd instar	*8*
4th instar	*8*

| *5th instar* | *8 (prior to cocoon construction)* |
| *Cocoon* | *300, if kept cool* |

The cocoons were placed in a plastic container with some large air holes and set in the vegetable crisper drawer. They were removed the following spring but only the males emerged, some dehydrated. Only one dehydrated female came out, which died the next day. Though there are other causes for wing deformities, dehydration prevents wings from expanding fully. The rest of the females looked as if death was due to being kept too dry, which was strange since the bulkier females should be less prone to desiccation. It was especially disappointing because I did not expect to see another wild promethea.

The next year in mid-July I walked outside during work (I rarely leave the building) and found a female promethea on the ground a dozen feet from a streetlight by the river. The same food plant and rearing techniques from the previous year were followed. However, during refrigeration, cocoons were misted well every few weeks instead of lightly every few months. This seemed to have been the key, since all but a few eclosed perfectly.

The moths emerged in a screen cage and a few mated and produced viable eggs. Hand pairing was attempted for two unmated pairs. Although artificial pairing was quicker and easier than with other species, both pairings resulted in infertile eggs. If it had been necessary to attract wild males, it was thirty days too early. Skewed emergence times are a pitfall of the abrupt temperature change caused by using a refrigerator instead of outdoor storage.

The fertile eggs resulted in my first captive-bred promethea caterpillars (versus previous captive-reared eggs laid by wild-caught adults). They grew

Callosamia promethea 3rd instar caterpillars

Callosamia promethea 4th instar caterpillars

well, emerged, and mated successfully the following summer. The following generation, and many subsequent, were maintained by attracting wild males to mate with the females. With multiple outcrossings, very little of the genetics from the 2005 female were likely retained, but the caterpillars photographed in 2018 carry some of her genes.

First and second instars are easily kept on food plant sprigs kept fresh in water containers. As usual, it is important to make sure caterpillars cannot walk down the stem to a drowning death. Place the container in a box—the box can be only a few inches high as long as the leaves do not hang onto or close to the edge. The box should have plastic packing tape—clear or brown—around the entire inside of the top. A thin layer of petroleum jelly is spread on the tape to prevent escape (the tape is necessary because petroleum jelly spread on a rough surface such as cardboard will not prevent climbing). They normally cling tightly to the foodplant, but if the food is not replaced before it is consumed or withered, the caterpillars leave. No matter how low the humidity, never mist the caterpillars or they will die. Third instar caterpillars can be moved to a larger setup of the same type or to sleeves outdoors.

In the wild, promethea feed on wild cherry and spicebush. It is sometimes called the "spicebush silkmoth." Alternative foodplants include birch, lilac, privet, *Prunus*, sweet gum, sassafras, and tulip tree.

Cynthia Moth
Samia cynthia (Drury, 1773)

The exquisite cynthia moth was introduced to the U.S. in the late 1800s, supposedly as a source for commercial silk. It is a member of the Saturniidae family, giant silk moths, which include various moon moths, Atlas, cecropia, and a few hundred other gigantic and colorful species. Atlas moths in the genus *Attacus* are often considered the closest relative of cynthia due to a number of similarities. The true silkworm moth *Bombyx mori* is from a different family, the Bombycidae.

In the United States, *S. cynthia* is found only in urban areas where the host tree, *Ailanthus* or Chinese Tree of Heaven, is established (the natural food of Atlas is also *Ailanthus*). However, in captivity *S. cynthia* can be fed other plants including willow, *Prunus*, lilac, and privet. Golden privet *Ligustrum x vicaryi* is a common landscape bush with thick, glossy, greenish-yellow, semi-evergreen leaves. Standard privet, *Ligustrum vulgare*, loses its leaves early in the fall and is not accepted by the caterpillars. Adult moths are often tiny if reared on willow, but that may be the result of cut leaves withering so rapidly. Caterpillars favor golden privet, which is easier to use than *Ailanthus* for the first three instars. Cuttings stay fresh longer and easily stand upright in a film capsule. The lower leaves of the cutting and/or paper towel should fill the space between the stems and capsule opening to protect the caterpillars from drowning. The young stay on fresh leaves, but if the leaves begin to dry out their behavior changes from sedentary feeders to restless wanderers. *Ailanthus* is useful in the last two instars when the caterpillars feed voraciously. During the last instar they eat more than 80% of all the food they eat in their entire life. A few dozen fifth instar caterpillars are capable of consuming a huge pile of leaves in a matter of hours.

Larvae are extremely intolerant of dampness and water. Feeding produces a lot of frass that can harm the larvae if it does not dry out. Larvae are usually reared in open trays, screen cages, or paper bags to keep the frass dry. Open trays must have a barrier around the top—petroleum jelly or Teflon™ paint—to prevent escape. Frass should be removed each time new leaves are added, but is not a danger if it is dry. Caterpillars should never be misted no matter how low the humidity. It is important to make sure water containers cannot tip over and spill, especially if kept in glass cages.

There are five larval instars, which is usual for moths. Hatchlings are black and turn light green with stripes after the first molt. Each instar has a unique color pattern. Following the second molt they cover themselves in a waxy white powder like Atlas moth caterpillars. Caterpillars appear primarily blue and white from instars three to five.

Before each molt they spin a silk pad and become motionless for a day or two. They will not move and can be buried if leaves are constantly tossed in. The silk molting pad of small larvae may be difficult to see but is extremely important. Without a molting pad they cannot extract themselves completely from their old skin. If caterpillars are accidentally removed from the molting pad they can be placed on a used pad or an old cocoon and they will molt perfectly.

STAGES IN DAYS AT 80°F (27°C)

Egg	*12*
1st instar	*6*
2nd instar	*4*
3rd instar	*4*
4th instar	*5*

| *5th instar* | *7* |
| *Cocoon* | *10-30* |

When fifth instar caterpillars attain full size, they begin to swing their heads back and forth as they spin silk. They silk a few inches around a leaf stem and up the branch, or make a large flat mat on the side of the cage, before they set to the task of building the spindle-shaped cocoon. They are designed to hang from a tree after all the leaves fall. They are able to survive dryness that would kill saturniid cocoons such as Luna or Polyphemus. Pupae should be kept in a vented container on a layer of paper towel. The paper is wetted when it dries out. If kept too dry they can die or become adults with crumpled, deformed wings. Cocoons are small in relation to the size of the adult moth. Atlas cocoon structure, size to moth ratio, and hardiness is similar. Pupae can be refrigerated up to a year. Unless food is available year-round, they should be placed into forced hibernation three to six days after the cocoon is completed. Otherwise, adults emerge in as little as ten days following cocoon formation. If caterpillars are reared at temperatures below 76°F (24°C) the pupae usually will not eclose without a three-month or longer cold period. Even if reared at 87°F (31°C), a small percentage of the cocoons may refuse to eclose without cold diapause. Low humidity can also slow down or prevent eclosion. Photoperiod does not seem to be an important factor in captivity. Adult males are smaller and have slightly larger antennae. The size difference is much more noticeable than the antennae difference. Small cocoons (from the same batch) are almost always the males. Adults are easy to sex because the male's abdomen is proportionately half the volume of the

Samia cynthia caterpillars

Samia cynthia cocoon

Samia cynthia
(CC Hectonichus)

female's. Males tend to have shorter wingspans but sexing by wingspan is not always accurate.

Females begin to lay eggs within 48 hours whether or not they have mated. It is best to save a dozen cocoons because they may not all come out together and the adult moths only live a few days. Males do not always mate with the females even if they come out together. It is normal to have a number of infertile eggs. Only one mating a generation is needed since even a runt female lays close to 100 eggs, while a nice-sized female can lay more than 200.

Cynthia moths are beautiful creatures whose expansive wings with soft pink and white tones rival any butterfly. The huge blue-and-white caterpillars are also very handsome. It is one of the easiest giant silk moths to keep in captivity and the specimens pictured (2005) are the sixth consecutive annual captive generation. Inbreeding has had no effect on adult moth size. The last two generations produced some of the largest individuals at 5 1/4 " (13.3 cm). The size increase was due to husbandry improvements including use of privet for early instars, rather than selection for size.

ATLAS MOTH
Attacus atlas (Linnaeus, 1758)
This is the king of all moths and is best known for having the largest wing area of all the world's Lepidoptera. This is also one species every beginning silkmoth rearer hopes to work with and cocoons are widely available in Europe. There are a number of subspecies found from India to China and southeast through the Indo-Pacific.

Attacus atlas

The Atlas moth is one of the few saturniids that can be easy to rear through multiple, consecutive (without the introduction of wild specimens) generations. Captive fertilization is often the sticking point for many lepidoptera. *Attacus* moths mate in relatively small cages and do not require flight time. The difficulty for temperate keepers is finding adequate food during winter.

If the cocoons are kept relatively humid, the adults eclose five or six weeks after the cocoon is spun. If reared together and maintained humid, nearly 100% will emerge within days of each other. Unassisted mating is common. This tropical species undergoes diapause in the cocoon when maintained under low humidity. If the cocoons are allowed to dry (or were purchased in diapause) it can be difficult to get a male and female to emerge close enough to mate successfully, even with a dozen or more cocoons. Also, if kept excessively dry, the survival rate can be low and surviving adults can have crumpled wings. Adults with imperfect wings rarely mate without hand pairing and hand pairing is only moderately successful.

The eggs are 1.5 mm across and pure white, but they are painted with a secretion during oviposition that results in a reddish-brown band around the outside and some lighter mottling on top. They hatch in less than two weeks.

Caterpillars are a translucent greenish-blue, but they are covered with a white, waxy secretion. The upper portions of the body, especially the long dorsal tubercles, appear to be coated in tiny snowflakes and the coating can be wiped off. They normally go through five instars and take about thirty days from hatching to producing silk. The fifth instar caterpillars are massive, but only about the size of a large cecropia caterpillar. The cocoons can be

Attacus atlas 4th instar caterpillar
(showing powdery coating on tubercules)

smaller than a cecropia's, even though the adult moth has twice the wingspan. The cocoon is a spindle shape and they do not create a bag-like outer cocoon.

Polyphemus Moth
Antheraea polyphemus (Cramer, 1776)
This is one of the most commonly encountered giant silk moths found around lights in urban and rural areas in North America. It ranges from southern Canada to Mexico. The moths are mostly brown in color, but they are very pretty. The wings have pink and purple highlights, as well as a gigantic purple, black, and yellow eyespot on each rear wing. The wingspan is around six inches (150 mm). When reared in captivity, males and females from the same group are similar in size.

The eggs are creamy white with a brown ring around them that makes them look a little like tiny buckeyes. They hatch after about twelve days into brown caterpillars. Oak is a commonly available food that works well. After they have eaten for a few days, they transform to a bright, neon green. The first through fifth instars do not change a lot in appearance. They resemble the caterpillars of the Luna moth (*Actias luna*), which is a very different looking, pale green moth with long tails on the rear wings. The Luna is also very common in North America, but usually feeds on sweetgum.

After about five weeks taken to reach 5th instar, the caterpillars begin to spin silk. The cocoons are pill-shaped and relatively small. Like most saturniids, they can be kept refrigerated for successful eclosion the following year. Cocoons can be stored in deli-cups with a few pinholes and misted once or twice a month. If kept at room temperature they emerge about four weeks

Antheraea polyphemus

Antheraea polyphemus caterpillar

Antheraea polyphemus first and 2nd instar caterpillars

Antheraea polyphemus cocoon

Antheraea polyphemus pupa removed from cocoon

after the cocoon is formed. They can be hand-mated with some success. However, food will run out unless the keeper is from an area with live oak. Large saturniids eat an unbelievable amount of food in the last two instars. The specimens in the accompanying photos were from an accidental second brood that hatched in mid-November because I did not refrigerate the cocoons in time.

It is difficult to get siblings to mate successfully, but drawing in wild males is very easy if they eclose from May to June. I put a female in a critter cage in the garage (with the door open) and there is usually a male present by 5 a.m. I close the garage door, catch the male, and place him in the cage with the female and they pair with little prompting.

Oslar's Eacles
Eacles oslari Rothschild, 1907

Oslar's eacles are found in Arizona in the U.S. and south through much of Mexico. The specimens depicted were found in Madera Canyon in Arizona. Males and females arrive at light traps in numbers. Males were visibly smaller at around 110 mm; females averaged 130 mm. All of the moths were brown and resembled dead leaves in appearance. None were marked in the bright yellow and orange patterns often seen in pictures. Oslar's eacles are widely variable in color and the moths can look very similar to the imperial moth, *Eacles imperialis* (Drury, 1773). I run across a few imperial moths every year around lights, but always find males. The imperial moth ranges widely across the central and eastern U.S. to southeastern Canada. The pattern varies widely but the wings are usually mostly yellow.

Eacles oslari
developing ova

Eacles oslari hatching

Eacles oslari
1st instars molting

Eacles oslari early instars

Eacles oslari caterpillar
on right preparing for
uncommon supernumeray
molt to 6th instar

Eacles oslari pupae

Eacles oslari

Imperial moths are among the unusual giant silk moths that do not form cocoons. They are members of the subfamily Ceratocampinae, which includes the humungous regal moth (*Citheronia* spp.) and the beautiful, tiny rosy maple moth (*Dryocampa rubicunda*). The final instar caterpillars burrow in the ground like sphinx caterpillars. Also, the eggs of these genera are very large compared to the size of the moth's body. At 2 mm, *Eacles* ova are among the largest of all lepidopteran eggs. The eggs appear yellow because of the yolk within. The shells are see-through, so the development can be watched. *Eacles* caterpillars start out yellow and change to green or brown as they grow. They seem to always become brown under captive conditions. They can be fed pine needles (*Pinus* spp.), but *E. oslari* survival rate is extremely low if fed pine. Hatchlings readily feed on oak (*Quercus* spp.), whether deciduous or evergreen, and the survival rate on oak is almost perfect through the last instar. However, the gigantic full-grown caterpillars are sensitive to humidity and overcrowding. A screen cage and removal of the frass twice daily is a must.

I found two things very interesting about Oslar's eacles. First, about 1/3 of the caterpillars were smaller than the rest, and yet they ended up about the same size. The smaller ones went through a sixth instar (the other 2/3rds were mature at 5th instar and pupated). Second, the pupae seem to eclose based on an internal clock after about 9 1/2 months without any attempt to cool or dry them to induce diapause.

Eacles imperialis
(CC Andy Reago and
Chrissy McClarren)

Eacles imperialis
caterpillar
(CC Benjamin Smith)

Dryocampa rubicunda
(CC Christina Butler)

Dryocampa rubicunda
caterpillars

Citheronia regalis
caterpillar

Citheronia splendida
3rd instar caterpillar

Silkworm

Bombyx mori (Linnaeus, 1758)

This easy captive is among the most ancient domesticated animals. Silk has been produced in China for nearly 5,000 years. The oldest English language book published on insects (Mouffett 1599) is devoted to the silkworm. The moths are short-lived, while the larvae are the longest active stage of the life cycle. Caterpillars are spectacular for displays and insect petting zoos. While the caterpillars are huge and impressive, the moths are relatively small.

Bombyx mori is considered a separate species from its wild counterpart. There are a few different cultivars which may or may not involve more recent crossing with wild stock. The Chinese version has nearly solid white caterpillars, pale adults, and white cocoons. The precursors to European cultivars were smuggled out of China about 1500 years ago. The Spanish cultivar has gray caterpillars. This "tiger" or "zebra" version has dark gray bands, constructs lemon-yellow cocoons, and adults can have extra gray markings. My first generation all produced yellow cocoons, while their offspring produced half yellow and half white cocoons.

In this species the primary modifications for domestication seem to be behavioral. Caterpillars are a real joy for those familiar with rearing other lepidopterans indoors. They do not run away the minute the foodplant dries a little, and can be kept on a flat tray. The prolegs hold on tightly but let go with ease, almost never tearing or ripping. Once they have molted a few times they can go days between feedings with little negative impact on growth. This is partly because they do not wander around searching for food and using up every last bit of energy.

Bombyx mori with
Spanish yellow cocoons

Bombyx mori
with eggs

It is possible that the behavioral traits which seem to be a result of domestication are natural traits that simply allowed them to be easy to keep rather than the result of domestication. The caterpillars can be kept on a flat tray and do not go anywhere, which can be attributed to helplessness. Yet when branches are placed within reach, they readily climb until the last leaf is found. It seems more efficient than the constant fleeing behavior of other caterpillars. Larvae produce a bit of silk even when they are not ready to molt. This could be the main reason prolegs are not easily damaged, since it pulls on the silk rather than the proleg pads. The adults cannot fly, but there are many wild moths with flightless adults.

Females lay a few hundred eggs on the cocoon, in a paper bag, or plastic container. The ovum is the winter resting stage for this species. Fertile eggs will change from light yellow to gray-brown within a few days and should be placed in the refrigerator (infertile eggs stay yellow). Eggs hatch after a few weeks or enter diapause relative to the amount of day length the adult moth sees, not temperature. However, I have had eggs from sibling females in the same cage, under identical conditions, where the eggs of one hatch and the other's go into diapause.

The first time I tried this species I could not find information on over-wintering the eggs and they all dried out. They do not look any different after they dry out. In a refrigerator they desiccate if exposed and so should be kept lightly damp inside a plastic bag. A little mold is normal and will not harm the eggs. However, excessive fungus growth will suffocate and kill them. Closed glass petri dishes may also work well. After a few months the eggs can

Bombyx mori
caterpillar feeding tray

Bombyx mori
caterpillars follow food
but do not wander

Bombyx mori
variability

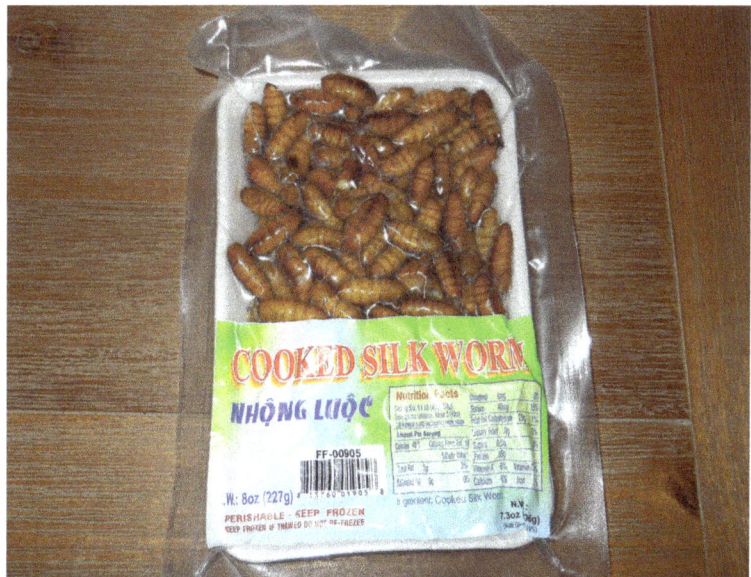

Bombyx mori
pupae are used
in several Asian
culinary traditions

be moved to room temperature for hatching. They can stay in the refrigerator a year or more, but the hatch rate declines after eight months. Eggs that do not diapause have a nearly perfect hatch rate.

Hatchlings emerge over a period of days to weeks, but will emerge close together if there is no diapause. They are extremely tiny and should be kept covered till after the first molt. They start well on hatching leaf buds since the buds retain moisture longer. The thinnest layer of standing water leads to mold which is closely followed by death.

The caterpillars become rather large and despite the small size of the adult will consume a volume of food equivalent to the large larval size. They reach three inches and can be larger than a promethea moth caterpillar (and eat just as much). They will nibble on many leaves other than mulberry, but cannot be reared on other leaves except possibly Osage orange. If leaves are not available, laboratory food mixes can be purchased from a number of sources (a component of the mix is dried, ground-up mulberry leaves).

Silkworms are large and impressive creatures despite the adult moth they will become. Even larger, more colorful moths are often reared for the handsome caterpillars because the adult lifespan is counted in days. Children absolutely love silkworms in handling displays. The caterpillars are commonly sold as nutritious food for insectivores (lizards mostly). They are a harmless prey items, but their lack of movement can help them evade many predators. Eggs and larvae of both the Chinese and Spanish cultivars are commonly available for use as food for other animals and are easy to acquire.

Dicogaster coronada (Barnes, 1904)

The unique development, interesting behavior, and massive size of *Dicogaster coronada* place it among the most impressive North American moths and yet it is rarely, if ever, kept. Despite the large number of moth guide books and some comprehensive rearing books I own, it is found in none of them.

I was checking out black light sheets in Madera Canyon which had an incredible variety and volume of moths including large sphinx and giant silkmoths. I noticed an extremely large, overly fuzzy moth, similar in size to the largest moth encountered on the sheets (*Eacles oslari,* which is larger than the local *Citheronia*). It was a gigantic moth, but was brown overall without interesting markings, so the size was its singular point of interest. Later I considered how amazing, huge, and colorful some caterpillars of small brown moths are and figured this species would likely have spectacular larvae. The caterpillar is the longest period of care and the stage I enjoy most. I grabbed a female from someone's sheet (nobody was keeping this one) and placed her in a brown paper lunch bag for egg laying.

The eggs took days longer than expected to hatch, but what hatched out was more of a surprise. They looked like little tent caterpillars, I was expecting some type of oakworm saturniid. I thought the moth might be a large *Anisota* species but it was easy to pin down to *Dicogaster* with the caterpillars.

The hatchlings showed more interest in eating the paper bag and would not touch the leaves. I wondered if oak (*Quercus*) was actually the correct food since every time I placed the caterpillars on a leaf they would wander off. More than half starved. However, after days of re-placing them on the

Dicogaster coronada
(CC Megan McCarty)

Dicogaster coronada ova

food and making sure no leaves touched anything they could crawl onto, they began to feed. They only ate for a few hours each night, grew slowly, and molted more slowly, but after forty days they had reached the final larval stage. The problem was at 5th instar they were the wrong color and barely an inch long—they would need to be at least three inches to become even stunted adults.

I have reared many dozens of giant silkmoths and sphinxes, other moths like silkworms, as well as many different species of butterflies. I have also read rearing reports on various species and had believed five instars was the rule for all Lepidoptera (the rare sixth instar occurring only in the odd individual of a few species (pers. obs. *Eacles oslari*)). Even the closely related tent caterpillars are known to have five instars (I had reared native tent caterpillars in the past). A caterpillar with a number of required supernumery molts was a big surprise, but how many?

They made it to 8th instar before I ran out of leaves. I tried some frozen oak leaves, but *Dicogaster* are very picky. Fortunately I had a friend with access to live oak.

This species is a member of the family Lasiocampidae, subfamily Lasiocampinae. The subfamily contains just a few North American genera: *Caloecia, Dicogaster, Gloveria, Malacosoma,* and *Quadrina. Dicogaster coronada* is the only member its genus. The genus *Malacosoma* includes that infamous and sometimes ravenous pest, the eastern tent caterpillar, *M. americana*. The various tent caterpillars are known for the massive web structures they build (although two of our six *Malacosoma* do not construct tents); the moths are small, brown, and plain.

Dicogaster coronada
early instar caterpillars

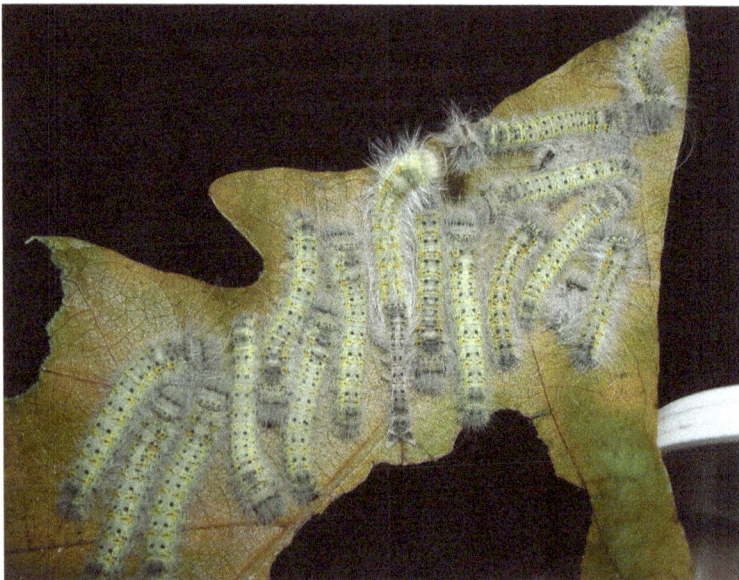

Dicogaster coronada
caterpillars

Dicogaster coronada
caterpillars

Dicogaster coronada
late instar caterpillar

Scientific names tell a story about the creature they describe, which can be informative, clear, or vague. The name *Dicogaster coronada* is informative and vague. "Dico" is from the Latin term meaning to "say" or "name" while "gaster" refers to the stomach or abdomen. The genus may have something to do with a "talking stomach" or dico- could be a misspelling of dicho- and possibly relate to the physiology of a forked stomach. The species name is more certain, but the story more convoluted. "Crowned" is the meaning of the Latin root of the Spanish *coronada* but instead refers to the surname of a famous Spanish explorer. The Coronado National Forest, named for this explorer, is located in southeastern Arizona and southwestern New Mexico and is a major part of the range of this moth.

Females lay a few hundred eggs, do not eat, and only live a few days. Hatching occurs after fifteen days. Molting intervals begin at once a week, but take longer for later molts. From egg laying, it took ninety-eight days for the caterpillars to reach 8th instar. 9th instar caterpillars formed cocoons in late December for a total developmental period of 150 days at approximately 75°F (24°C). The 'cocoons' were just a few threads loosely holding some leaves together. Moths emerged after thirty days (end of January). In nature the 9th instar caterpillar likely feeds much longer and slows down, with cooler temperatures leading to much larger size and a stronger cocoon; full-grown caterpillars have been found in June in the wild (McMonigle 2010).

As usual, the instars progress through different color stages. Hatchlings are white, the next few molts are light green with orange and black spots, then black with lighter bands and spots, and finally dark brown with white bands.

They molt as a group and are highly gregarious. Through 7th instar, larvae wait until the entire group has molted (which can take more than two days) before any begin to feed. Proximity to sedentary siblings prevents feeding. If individuals that molted first are removed and placed on new food, they will begin to feed immediately. Caterpillars hang out in a group under one leaf and travel after dusk to feed on nearby leaves. The group returns to the original leaf after a few hours of feeding and doesn't abandon this leaf even after it is long dead and dried. This behavior is less noticeable in the late instars because of increased food requirements. Also, they switch from congregating on leaves to branches at 7th instar.

Crepuscular behavior and hairy bodies are the primary defenses, however, they also possess setae that break off and penetrate human skin. Due to small size of the setae, this defense was not apparent before 7th instar. It is barely noticeable on the tough skin of fingertips, but might be painful on softer, thinner skin. The setae appear to lack venom, since the surrounding skin did not become inflamed or turn red.

Caterpillars are picky and have little ability to locate food unless it is touching the branch they are on. The natural foodplants of this species are live oaks (*Quercus emoryi* and *Q. oblongifolia*), but white oak (*Quercus alba*) is well accepted in captivity. Red oak (*Quercus rubra*) is refused. Caterpillars can be forced to eat pin oak (*Quercus palustris*) but they must be placed back on the foodplant a few times daily as they wander off until they learn to accept the food. Even when a readily accepted oak is used, if they wander onto the sides or bottom of the cage they must be placed back on the food because they will not find their way back. Although this species has a narrow acceptance of plants in the genus *Quercus*, some common food plants (hybrid privet, strawberry, pyracantha, leatherleaf viburnum) were offered to 7th instar caterpillars without even a test nibble. Texas red oak (*Quercus buckleyi*) was then offered and at first eaten with some reluctance (half had to be placed back on the cutting before they accepted it).

Dicogaster coronada is a fascinating animal with advanced behavioral adaptation that displays as oddly shy and communal caterpillars. The huge size and long growth period also make rearing interesting, but unfortunately they can't be reared in areas where evergreen oak isn't available and may be difficult to rear to full-size in captivity. The large number of instars, nine, seems to be an especially unique feature, though it is unlikely to be the only lepidopteran to exceed the usual five instars as part of its required growth.

Tobacco Hornworm Sphinx Moth
Manduca sexta (Linnaeus, 1763)

The monstrous caterpillars of this spectacular giant moth have become a common sight at reptile shows over the last ten years as a specialty food for large insectivores. Eggs and caterpillars have been available for classroom use from biological supply houses for decades. Kits have long been for sale which include artificial diet, pupation substrate (soil), rearing containers, and moth eggs. Tobacco hornworms are commonly found in the garden on cultivated tomatoes, but so are tomato hornworms (*Manduca quinquemaculata*). *Manduca* species are closely related to the famous death's head moths (*Acherontia* spp.) popularly cultured in Europe. While many moths are known to stridulate

Due to its diet, this laboratory-raised *Manduca sexta* is an eerie blue color

to make sounds that can ward off bats, death's heads make audible noises with their mouthparts that can startle humans.

When fully grown, the blue and white caterpillars are the size of a human finger. Caterpillars are usually cultured on artificial diet as the volume of leaves needed to grow large numbers would make rearing a near impossible chore. Frass should be removed daily and food should be constantly available. The most common diet has blue dye, but caterpillars that eat it become aqua blue because the diet lacks carotenoids. They are green if fed the leaves of European nightshade (potato, tomato, tobacco, etc.) or grayish green when fed potato slices.

When mature they begin to shrink and crawl around the bottom of the cage as they search for a place to pupate. In nature, the caterpillars burrow to form an underground cell compacted by the wriggling movements of the larva. Silk is not used to reinforce the sides of the cell. A shallow layer of damp substrate is adequate in captivity since the pupae do not require protection from predators. The distinct pupa looks like a bottle with a handle and the "handle" contains the developing proboscis.

After a few weeks the adult moths emerge and within a few days they will grow hungry. The moths normally only feed as the sun is going down and only uncoil the long proboscis in mid-flight which requires a very large cage in captivity. Alternately, the proboscis can be carefully unrolled with a tooth-pick and placed in sugar water. It takes about five minutes to feed a single moth one time. It is important not to damage the proboscis. Even with a good feeding schedule, moths only live a week or two. Females are around thirty

Manduca sexta
1st instar caterpillar

Manduca sexta
caterpillar molting

Manduca sexta
freshly molted to pupa

Manduca sexta
pupa after eleven hours

Manduca sexta
pupa at eighteen days

Manduca sexta
moth feeding

Freshly emerged *Manduca sexta*
adult moth with meconium

percent larger than males. Adults mate readily and females lay hundreds of eggs to restart the life cycle.

Though a neat animal and good specialty food, the amount of work required to rear tobacco hornworms for personal use sets them well beyond the grasp of most reptile and large insectivore keepers. Additionally, while large batches of medium can be made cheaply with available recipes, pre-made medium is very expensive to purchase. Other drawbacks include caterpillars' constant need for attention, including frass removal and the unpleasant odor of the artificial diet (sort of like baby poop).

The handsome moths have a three- to four-inch wingspan (up to five inches in the wild). The wings are gray while a row of six (hence the species name *sexta*) large yellow spots runs up each side of the abdomen. Tobacco hornworms are a great treat for large insectivores. It's enjoyable and takes little effort to purchase a group of full-grown caterpillars and observe their development into moths.

White-lined Sphinx
Hyles lineata (Fabricius, 1775)
The white-lined sphinx is one of the largest and most common sphinx moths in North America. Although adults are most commonly active at night, individuals can be seen feeding from flowers in the early morning—rarely during the day. They have a long proboscis and feed from large, tubular flowers like honeysuckle and columbine. The food preference and approximate size of the moth result in another common name, hummingbird moth. The large and impressive caterpillars are variable in coloration. The base color can be green, black, or yellow. In contrast, the moth coloration is very consistent.

The most common way for a collector to find specimens is to search the sides of buildings near lights, before sunrise. I usually stumble across two or three adults each year from mid- to late summer, usually from July to early September. In 2018, however, I was surprised to run across nearly a dozen over the first two weeks of October. When I first encountered three specimens the same morning, I imagined the unseasonably warm temperatures caused some to emerge from their pupae early rather than overwinter. A few moths and a week later I remembered we had planted a large number of spider flowers (*Cleome* sp.) along the front of the building. These flowers were likely the only available source of food the moths could locate so late in the season. Large sphinx moths require a significant source of nectar.

I collected a few and tried to feed the moths with red beetle jellies with some success. Yellow and orange jellies were offered and ignored. Each color is a different flavor and odor, so the attractant may not have been color. I observed a few moths feeding in the early morning as pictured, but they are nocturnal, so the primary signs of feeding were swipe marks across the tops of the jellies made by the moths' proboscis. Despite eating successfully, the moths did not seem to have much life left in them.

I often run across several different sphinx moths, including the larger *Manduca* and pandora moths, every summer. However, I only seem to find males attracted to lights. Ten of the *H. lineata* were males, but there were also two females in the group. The females laid a few dozen tiny green eggs on the last days of life.

The eggs are neon green, flattened spheres. Ovum size (barely 1 mm across) relative to the rather massive abdomen of the female sphinx moth is surprisingly

Hyles lineata
egg and 1st instar
caterpillar

Hyles lineata
1st and 2nd instar
caterpillars

Hyles lineata
3rd instar caterpillars

Hyles lineata
4th instar caterpillar

small in relation to ova of other moths and butterflies. Wild grape leaves and European nightshade cuttings were provided to induce oviposition, but no eggs were laid on them. The eggs were mostly laid on wood and plastic surfaces (one was dropped on the soil). Ova were moved to a leaf's surface to prevent desiccation. Some were damaged as they were removed from the wood because they were firmly glued in place. Once removed, they could be handled safely with fingers (versus delicate microlepidotera eggs that should be manipulated with a soft, paint brush).

STAGES IN DAYS AT 75°F (24°C)

Egg	*4-5 days*
1st Instar	*6 days*
2nd instar	*5 days*
3rd instar	*4* days*
4th instar	*6 days*
5th instar (feeding)	*7 days*
Prepupa (5th)	*4*
Pupa	*18-19 days (up to 300 refrigerated)*

*Usually later instars require progressively more time to grow and molt, so this growth pattern may have been related to a less desirable food.

The apple-green 1st instar larvae ate holes in the middle of leaves and jumped from leaf to leaf. This sporadic feeding pattern of eating holes here and there continued through 3rd instar. 2nd instar transformed from green to

Hyles lineata
5th instar caterpillars

Hyles lineata
freshly molted caterpillar

Hyles lineata caterpillar with
primarily green coloration

black with a line of yellow dashes down each side. 3rd instar looked similar to 2nd, but the head turned brown and an additional yellow line developed on each side, down near the prolegs, along with clusters of small, white spots in-between. At 4th instar the background color turned back to green, with black along the top. Some retained brown heads and more black on the body, while others had green heads and less brown and black on the body.

The hatchlings refused to eat nightshade but fed on the grape leaves. I was not expecting the nightshade to work, but since this species can eat tomato it was worth a try because nightshade cuttings can stay fresh indefinitely and grow roots. Grape leaves are a difficult food source that wilt rapidly. The Virginia creeper leaves had already turned red for the fall. Even when grape cuttings are placed in water, the leaves begin to dry up in a day or two. In order to preserve cuttings I sprayed the leaves with a powdered milk and yeast solution (3 oz. water, 1 tablespoon powdered milk, 1/4 teaspoon baker's yeast). The leaves should be allowed to dry (stems still in water) before offering as food. The coating seals moisture inside the leaves and can improve survival of small caterpillars. As the grape leaves outdoors turned mostly yellow, I added a branch of golden vicary privet (a semi-evergreen hybrid *Ligustrum* that holds its leaves until they freeze off). One freshly molted 3rd instar moved to the privet without prompting. It was a molt behind its siblings and probably was looking for some peace more than trying to find a new food. A fourth instar later moved over on its own but was found on the bottom of the cage a few times and was placed back on the privet and then the grape. All were fed on privet for most or all the fifth instar. Some of the privet had turned purple from below freezing temperatures and was offered to the last 4th instar

Hyles lineata
ready to pupate

Hyles lineata
pupa ready to eclose

Hyles lineata
moth feeding

Hyles lineata
older moth

in hopes it would become mostly black when it molted to fifth instar (larvae of this species can be variably colored). Instead, it was the only specimen to lose all its black striping.

The fifth instar is a large caterpillar (76 mm or 3″) with a relatively small head (4.5 mm or 3/16″). When full grown it stopped eating and crawled around the bottom of the cage. Within a few days it lost 1/3 of its length and 2/3 the mass. It burrowed under objects and flipped over the water containers for the branches. The wandering stage should be moved to a shallow container of dirt for pupation. Care should be taken as it becomes an avid biter at this time. The bite is not painful but surprising and could lead to a dropped specimen. Each larva should have its own container and should be left undisturbed. It builds a loose earthen cocoon and lines the walls with silk. If disturbed early on it will rebuild and use more silk which can result in a smaller adult. A few more days and the larva molts to become a pupa. These can be placed in the refrigerator in damp soil and removed for eclosion in late spring.

The white-lined sphinx is an interesting captive. Caterpillars are easy to raise indoors and like most sphingidae the caterpillars can survive without excessive ventilation. I have yet to learn how difficult it is to mate adults in captivity.

WAVED SPHINX
Ceratomia undulosa (Walker, 1856)
This is a medium-sized hawk moth with a wingspan of three to four inches. Adults are difficult to see on the side of a tree because they resemble gray,

Ceratomia undulosa
(CC Andy Reago and
Chrissy McClarren)

Ceratomia undulosa
caterpillar

weathered bark. Of course, they are easy to pick out on the sides of buildings. There is a round pattern on the pronotum that very vaguely resembles the pattern on a death's head moth. A close relative, the catalpa sphinx, *Ceratomia catalpae* (Boisduval, 1875), has similar-looking moths but fantastic yellow and black caterpillars. I collected a female *C. undulosa* in 2007, just before dawn, from the side of a building, beneath a light. She laid a few dozen eggs and died without feeding. I reared six consecutive generations on golden privet. The first generation did not seem to feed well and about half died. However, the following generations readily accepted the privet and experienced nearly perfect survival.

Caterpillars are brightly colored, neon-green with pale stripes. They go through five instars and feed on ash, oak, lilac, and privet. The final instar is the most colorful, with a two-tone green body and red markings behind the pale stripes. About the time it stops feeding and begins to shrink, the top of the caterpillar turns red. This stage can be moved to a shallow container of damp dirt for pupation. The caterpillars do not produce any obvious silk and use wriggling movement of their bodies to carve out a pupal cell in the soil. Pupae can be overwintered in a refrigerator and placed in damp paper towel or soil in a container with a few pinholes for ventilation. Adults usually only live a few weeks and do not easily feed in captivity. However, they mate and lay eggs without feeding. Mating is easy and inbreeding over multiple generations is not a problem (like silkmoths). I did not make it to the seventh generation because I was too late setting up the pupae for refrigeration and ran out of foodplant.

Ceratomia undulosa
molting to 3rd instar

Ceratomia undulosa
caterpillar turns red
when ready to burrow

Ceratomia undulosa pupa

Ceratomia catalpae
caterpillar

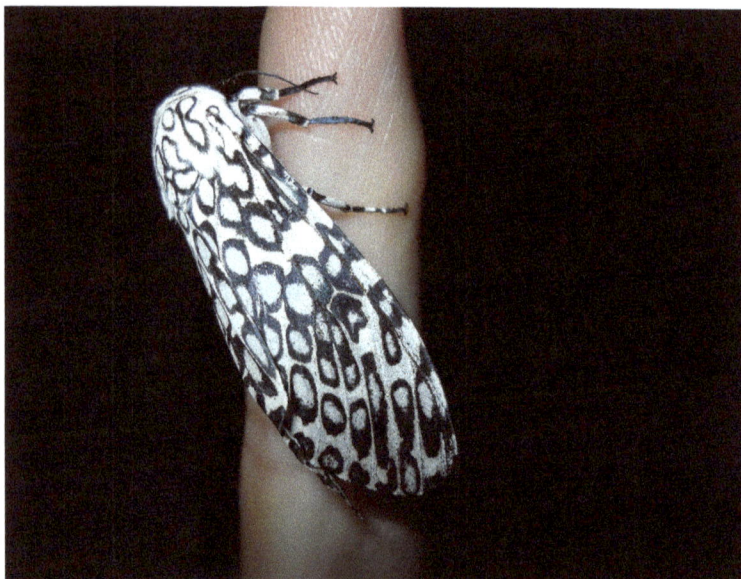

Hypercompe scribonia

Hypercompe scribonia
(CC Andy Reago and
Chrissy McClarren)

Giant Leopard Moth
Hypercompe scribonia (Stoll, 1790)
In early September of 2012, I came across two caterpillars of the giant leopard moth (*Ecpantheria scribonia*). Each was walking across a cement floor at my work, ten days apart but not more than five feet distant. When I walked into the second one I searched around the building, exterior vegetation, and rocks but could not find any more. I knew what these caterpillars were immediately, although I had never seen one. The brightly colored moth is gigantic for a tiger moth and the first and only wild adult I had encountered was noticed on a wall outside work a few years earlier (unfortunately a male). It was snowy white with large spotted wings and deep black circles on the body. The legs are black and metallic blue. Alternating metallic blue and orange markings run down the abdomen. Before this time I did not know my area hosted such a large and brightly colored moth beyond the usual saturniids and sphinxes. There are some giant underwings, but they look like a piece of bark unless the wings are open. The beauty of this "new" giant moth stuck in my mind so I looked up and remembered the appearance of the caterpillars. These massive woolybears could have been nothing else.

Like the common *Isabella* woolybears, *E. scribonia* caterpillars have long, stiff setae and they curl up into a spike-covered, irregular ball when disturbed. Unlike other common tiger moths I have kept, the setae do not break off pieces during handling. The setae are much thicker in cross-section, and less numerous. The red-banded, black body is visible through the sparse setae. I placed the large woolybears in a screen cage. Although full-grown Arctiid larvae are usually fine without food for some time, I was not sure I could find something they would feed on so late in the year. The first caterpillar turned out to be fond of semi-evergreen *Viburnum rhytidophyllum* leaves. The second woolybear was extremely thin for a fifth instar so I was fortunate to have already stumbled upon an acceptable food. This viburnum keeps some of its leaves though the winter and the caterpillars were kept near the floor (at around 68°F (20°C)) so I imagined they would feed on and off through the winter.

The huge woolybears would dig into a layer of dried leaves and dry frass (from other caterpillars that summer) for a day and then and feed on the fresh leaves for a day or two. After a month they had not surfaced for some time. I dug though and found large, loosely formed cocoons. These were placed in a 32 oz. deli cup with no ventilation and no moisture, since the leaves would mold and cause suffocation while vents would certainly desiccate the pupae if they did not hatch till spring. Some moths (like daggers) wait till spring even if kept at room temperature, but just in case a plastic ladder was placed in the deli cup for emergence (this is also why the tall deli cup was chosen). The "ladder" was cut from plastic yarn canvas. After another three weeks I took the pupae out of the cocoons and, by size, was certain they were male and female. The male cracked along the wings when handled, but it did not bleed. He emerged perfectly by the next morning. The female seemed a bit behind, but she emerged five days later and the male was still alive. He was found hanging around the female, but no mating was observed.

She did not lay eggs after a week, so I placed a small viburnum leaf in to promote egg laying. I cannot say she would have died without laying eggs, but that very evening she began to lay small silver globules. The eggs were

Hypercompe scribonia
with eggs

Hypercompe scribonia
eggs developing

Hypercompe scribonia
caterpillar
(CC Kerry Wixted)

Hypercompe scribonia
cocoon

surprisingly small. The body of the female was as massive as all but the largest of our giant silkmoth (native saturniid) females. However, her eggs were smaller than painted lady butterfly eggs. I did feel a desire to not count each one, but I would estimate close to a thousand eggs were laid. After three weeks a number of the eggs started to turn pink in color. I was excited since color changes in eggs usually indicate late stage development. Despite the eventual size of the caterpillars and difficulty of acquiring winter food, I think I had enough viburnum to feed a few dozen to adulthood. However, after another week or so it seemed none were going to hatch and the color change may have been related to dying (the extremely dry air due to constant running of the furnace may have been a problem even though ventilation was severely restricted to reduce desiccation).

Milkweed Tussock Moth
Euchaetes egle (Drury, 1773)
This lepidopteran ranges across the eastern United States, up to Quebec and Ontario, Canada. As far as anyone knows, the species name *egle* does not mean anything, while the genus means new and "hair-like," possibly since they are hairy caterpillars from the New World, though new or good might reference anything. The common name is also descriptive of the caterpillar rather than the moth. As both the genus and common name suggest, the impressive quality of this species is the larval stage. The caterpillars are memorable, but more so when found in huge, brightly-colored groups. The bright orange and black caterpillars are outfitted with long tufts of hair-like setae that lead to the common name tussock (a tussock is an area of grass that is longer than

Euchaetes egle
(CC-SA Patrick Coin)

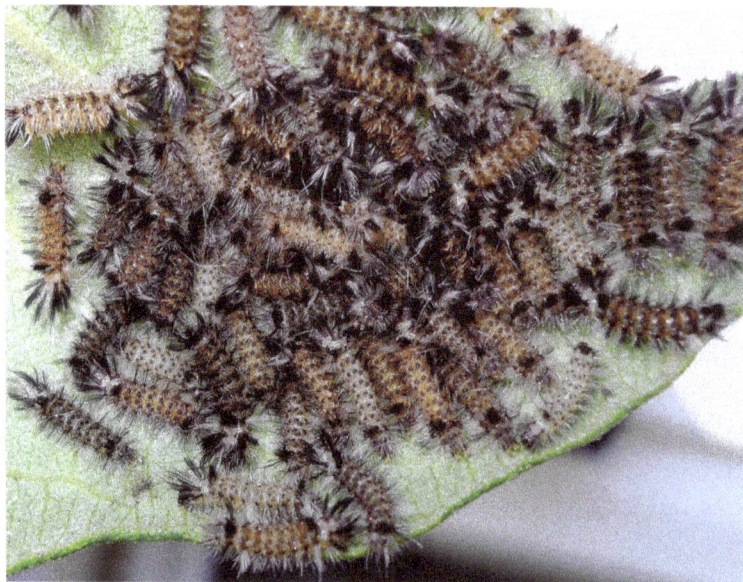

Euchaetes egle
2nd and 3rd instars

the grass growing around it). As the other half of the common name suggests, larvae are primarily restricted to feeding on milkweed, *Asclepias* species.

The adult does not feed. This tiger moth is somewhat drab and small, yet still is larger and more colorful than the average moth. The wingspan ranges from an inch to an inch-and-a-half, with females being the largest. The wings are an unusual soft gray (like baby blue, if it were printed in black and white) while the abdomen is orange with rows of black spots and white bands. Males are a little smaller than females and buzz their wings for mating like silkworm moths. Captive-reared adults mate readily in captivity but do not produce many eggs, and only on fresh leaves.

The eggs are incredibly tiny and laid in small rafts on the leaves. Indoors, a non-vented container must be used for gestation and early development if a live plant is not kept, otherwise the leaf and eggs rapidly dry out. Hatchlings are green for the first instar and yellow the second. By the third they begin to develop the familiar form of a tussock caterpillar. By the fourth instar they should be moved to a vented container or moisture from the frass will cause them to suffocate and die.

Outdoors, the caterpillars are found in tight-knit groups until the last instar, or earlier, if they disperse because they have completely stripped the milkweed plant they started on. They live in fields and disturbed areas along with the host plant. Many probably die from running out of food since the female moth does not make any attempt to determine the number of plants in the area and larvae are very restricted in food-plant options. Sometimes, when there are plenty of leaves, the caterpillars abandon the host plant and can be found under dead leaves and large rocks within a few feet of the host.

Euchaetes egle caterpillars

This may result from a predator disturbing them, but there are few large predators that will eat these chemically and physically protected creatures. When hiding, they curl up in a circle and are difficult to pick up. The hairlike tufts break off when grabbed and during handling. These can cause great discomfort in the soft areas of the face. Caterpillars have little protection when really tiny and at this time a single web spider can eat an entire brood. Spiders do not present much danger after 3rd instar, possibly due to the elongate setae rather than increased effects of chemical protection.

Cocoons are formed with thick, protective silk—the long, defensive urticating setae are incorporated in the silk matrix. Inside, larvae molt to become pupae in about ten days and are easy to refrigerate successfully. Unlike most lepidopteran pupae, they are not sensitive to a little dryness. I put them in the refrigerator a month after they pupated, hardly checked on moisture levels, and left them in three months longer than I should have and they emerged fine (any one of these errors would have resulted in total failure for much of anything else). I was impressed to see any, let alone every, moth emerge successfully.

The moths fly from May to September and there are two generations per summer in much of the natural range. Caterpillars are found beginning in June. In captivity the adults only live a week or so.

Euchaetes egle is a sturdy tiger moth that is interesting to cycle in captivity. Caterpillars make an eye-catching display, though only in the last two instars, which lasts only a few weeks. I have not experienced irritation caused by the setae, but I have been careful not to wipe my face after touching them.

MEXICAN JUMPING BEAN MOTH

Cydia deshaisiana (Lucas, 1858)

The jumping bean is a common novelty toy, but the item seen in stores is usually an inanimate piece of brightly colored plastic with a ball bearing inside. The real Mexican jumping bean is 1/2" (12.7 mm) long, brown, and contains a living creature that moves on its own. Most people in the United States and many around the world have heard of the extraordinary jumping beans of Mexico. However, most people under thirty have never seen a real jumping bean. My parents bought me some a few times from a local five-and-dime store in the 1980s, but I hardly remember them. I finally acquired some again from a curio shop in the Phoenix Airport on the way back from the ATS Conference in 2009.

The female jumping bean moth lays her eggs in the flower of a Mexican shrub. The shrub, *Sebastiana pavoniana*, grows on rocky slopes in the Mexican states of Sonora and Chihuahua. The 'bean' is actually a carpel—section of a seed capsule. As the caterpillar grows it becomes large and heavy enough to move the carpel when it twists its body. The carpel does not really jump, but shakes and rocks back and forth (apparently, Mexican shaking carpel was not a catchy name). The caterpillar consumes the entire seed as it grows and leaves only the thin, hard shell. Inside this strong fortress it changes to a pupa and then exits as it molts to maturity through a circular hatch it made as a caterpillar. The bean only moves when the caterpillar is nearly mature, so the shelf life is a few weeks to a few months. Warm temperatures speed up development. Seasonal crops have been cultivated and harvested for decades but there is no record of indoor captive breeding attempts, because the adults are short-lived and would have to be timed with live, seeding plants. Interestingly, another very famous caterpillar—the proverbial 'worm' in the apple—is also from the genus *Cydia*.

Jumping beans

Jumping bean moth
caterpillar and pupa

Danaus plexippus,
Monarch butterfly

Danaus gilippus,
Queen butterfly

BUTTERFLIES

Monarch Butterfly
Danaus plexippus (Linnaeus, 1758)

No matter what part of the world you live in, you are probably familiar with the monarch. This incredible butterfly is found on every continent and most islands where milkweed grows. The monarch flies great distances across land and sea and colonized most North and South America by its own power eons ago. Few other arthropod species could compare. Only over the last few hundred years, with man's ability to travel globally have some arthropod hitchhikers been able to enlarge their range as expansively as the monarch.

The monarch butterfly is not only famous for its range, but also for its intense migrations. In North America, adults fly as far as two thousand miles each fall to spend the winter in a few cool mountains in Mexico. In early fall, groups of a few dozen to thousands can be seen resting on branches across the U.S. and Canada as they make their way down to their winter stomping ground. The large groups are an amazing site but cannot compare to the final stop where every tree seems to be covered in leaves made of butterfly wings.

Adults are large butterflies with a wingspan of four to five inches (10-13 cm). They are magnificently colored in deep black and orange and appear nearly identical wherever they are found. Remote island populations are relatively isolated, with the lone straggler from the mainland only showing up every few hundred, or few thousand years. Roughly ten percent of the monarchs found on the Hawaiian Islands are white where they should be orange.

The full-grown caterpillar is about 2" (5 cm) and as magnificently colored as the adult. The body is circled in alternating bands of yellow, black, and white. Both the front and rear have a pair of long, black tubercles resembling antennae. Caterpillars of the queen (*Danaus gilippus*) look similar, but have a third set of long tubercles just after the thorax.

I remember keeping monarch caterpillars as a class project in elementary school. I cannot recall the grade or teacher, but it was probably second or third grade. What I do vividly remember is the look of the strange caterpillars and the neon-green, barrel-shaped chrysalises that transformed to black and orange as the butterflies inside developed. I remember searching numerous patches of milkweed plants that summer and many summers afterwards for caterpillars. Unfortunately, I live in a cold, distant area of their range and never found any.

Fast-forward some decades. There is a large scrub area located on the northern edge of my property. Milkweed, willow, wild rose, blackberry, elderberry, and many other plants grow there. I cut back nearby plants to help the milkweeds grow. I like to observe the colorful milkweed longhorn beetles and large and small milkweed bugs that live on the plants. The milkweed also provides giant, spherical clusters of small purple flowers that smell sweet like orange blossoms. However, most of all, I hope a monarch will find the plants acceptable for egg laying.

A few summers ago I was outside and noticed a monarch paying special attention to my milkweed plants. From thirty feet away, I watched her flying and landing for nearly an hour. It was fun to watch her, especially since I do not see such large butterflies very often. After she left I spent half an hour closely inspecting the leaves of dozens of plants. The eggs are incredibly tiny (about a 30th of an inch, <1 mm) and as green as the leaves, but not too difficult to locate because they are elongated and project off the surface. She had carefully attached seven eggs to seven leaves, no more than two on a single plant.

I was glad to see my milkweeds had finally attracted their intended target. I decided I would leave the eggs where they were so they could be observed as they developed naturally. Then I could easily walk out and check their progress each day without harming them. After a few days, the eggs began to darken as the caterpillars developed inside. I believe they hatched before being eaten, but I will never know for sure, nor do I know what ate them. Sometimes I forget nature is hungry and more likely to kill things than I am.

A few years later (2003) I attended a family picnic a few hours south of my home. The picnic was at my uncle's, on a former farm surrounding a one-acre lake lined with trees. After lunch I walked alongside the pond and noticed half a dozen meager milkweed plants growing next to the shaded bank. I performed my usual "monarch check" and found a one-inch caterpillar and two tiny ones (less than 1/8th inch or 3 mm). I decided not to leave these to be eaten. I was excited by my find and spent hours later that day searching hundreds of milkweed plants throughout the lot. I found many insects—including various predatory assassin bugs that might explain an absence of caterpillars—among the milkweeds in the fields. However, the only caterpillar I found was another tiny one on the original group of plants in the shade.

The milkweed plants in my yard finally had their chance to be food for monarchs, even though the four caterpillars only ate half a dozen leaves total.

I rear some native species of sphinx moths and giant silk moths each summer and I am always impressed by the caterpillars' rapid speed of growth compared to most other invertebrates. I was not prepared for the monarch's speed of growth. The caterpillars made the faster moths seem slow. In four days the tiny caterpillars were over an inch (2.5 cm) and the large one was mature at two inches (5 cm). By day six the oldest larvae had already transformed into a chrysalis. On day nine, the remaining three transformed. On day fourteen, the first adult butterfly emerged. The other three emerged on day seventeen.

I knew monarch butterflies were much larger than the usual butterflies in the area and had seen them flying around, but I was still struck by how huge they appear up close. All four adults had a wingspan right around five inches (~13 cm). They were even bigger than most of the exotic butterflies at the local zoo's butterfly exhibit and about the same size as many of the native giant silk moths.

I hoped to begin a continuous culture. Then I could enjoy them each year and not have to rely on the whims of the seasons. Maybe I could freeze milkweed leaves and raise caterpillars during the winter. Or, maybe I could get the adults to live till the plants emerged from their winter rest.

Danaus plexippus
caterpillar

Danaus plexippus
chrysalises

The only flight cage I could find was made of screen and measured 14″ x 14″ x 24″ (36 cm x 36 cm x 61 cm). I assumed that, like moths, they would quickly beat their wings into worthless tatters if I put them in a glass or plastic cage. I stood the cage on end and hung a hummingbird feeder inside. I filled the feeder with "hummingbird food" because it is formulated to approximate flower nectar. I hoped they would be willing to feed from the fake flowers. The cage was a bit on the small side but they seemed content to flutter from one side to the other all day long.

After a few days I began to worry. I had not seen them drink from the feeder. I also worried they might become restless and beat their wings off in the small cage. They never damaged their wings but after ten days I could not get them to feed. I decided to let them return to the wild. Maybe one or two would make it to Mexico in the fall.

Afterwards, I learned the common adult food-flower in my area is ironweed (J. Smolka). The flowers are purple. Hummingbird feeders are red, which may have been the problem.

Weeks later, I noticed a monarch inspecting my milkweed plants. I ran up (scared the butterfly away) and searched for eggs—no luck, maybe next year.

MOURNING CLOAK BUTTERFLY
Nymphalis antiopa (Linnaeus, 1758)
Nymphalis antiopa is found across Europe and Asia to Siberia and Japan. Its common name in most countries is the native language version of mourning cloak, though in the U.K. it is commonly called the Camberwell beauty. It is also native across the U.S. into Canada and south through much of Mexico.

Mourning cloaks, *Nymphalis antiopa,* feeding

Nymphalis antiopa
5th instar caterpillars

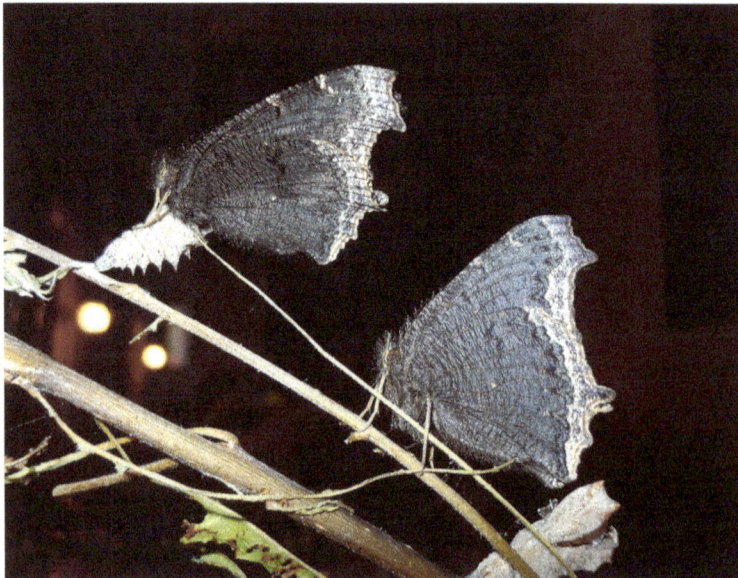

Nymphalis antiopa
adults, wings closed

It is the state insect of Montana. The adults migrate in winter, but the movements are unknown.

The caterpillars are commonly found on willows located on the shores of lakes or riverbanks. It can be impossible to collect them without a pair of deep waders since the females usually choose to lay eggs on branches overhanging the water. Indoors, willow dries out quickly even when freshly cut and placed in water. The food may have to be replaced daily as it dries out before being eaten. Rapidly withering foodplant is usually harmless, but can cause caterpillars to wander around the cage without eating much. Drying leaves cause mature larvae to transform to chrysalises before reaching full size. Withered leaves result in stunted adults that can be nearly as small as a large cabbage white butterfly. Other foodplants such as poplar, elm, and birch may be used but refusal and wandering from the food can cause worse problems than a little withering.

During the last two instars caterpillars have branching spines that can break off in human skin and cause a sharp stinging sensation (generally the spines are too small in earlier instars to be noticed). Usually only one or two spines enter the skin and the pain subsides in a few seconds but handling and cage cleaning can be uncomfortable.

This handsome butterfly is found across much of the globe but this species' mastery of the Earth's environment is not what makes it so special. It is an amazing lepidopteran because the adult stage is readily kept in captivity for months rather than days. The adults remain alive for 270 days at room temperature with limited care. Other butterflies and moths cannot find food placed in front of them (even if they have working mouth parts), destroy

Nymphalis antiopa caterpillar and chrysalises

themselves by flying incessantly, or have naturally short lifespans. Morning cloaks do not beat their wings to pieces in small flight cages and do not seem to be bothered by metal screen or square cages (circular cloth screen caging is better). The wingspan is usually around three inches (7.6 cm).

Adults must have access to food daily when they are kept at warm temperatures, less if kept cool. Butterflies can be provided nectar substitute or sports drinks, though I have only had good success using beetle jellies; the red, brown sugar-flavored, or banana-flavored varieties. It is very important that the feeding cup is next to a screen cage wall so they can walk down to feed. They begin to eat three to four days after emergence and will be seen feeding regularly if the food is accessible. Without access to food the butterflies will not last for many days. Daily misting with water is helpful as desiccation is the most common cause of premature death in captivity.

As usual for butterflies, their waste looks like little paint drops that range from pink to red in color depending on the food. Their waste does not smell bad or pose a risk to the animals. A shallow layer of substrate can be kept on the bottom and stirred up now and then in lieu of daily cleaning.

COMMON BUCKEYE BUTTERFLY
Junonia coenia Hübner, 1822
The unique and handsome wing coloration leads to the common name and makes this species easy to identify in the field. The adults are medium-sized butterflies with a wingspan from 1.8-2.8″ (45-70 mm). Their wing pattern reminds me of owl butterflies and morphos (the underside) at the Costa Rican exhibit at a local botanical garden. Although common across most of the

Common buckeye, *Junonia coenia*

United States, and ranging into southern Canada during summer, any adults that do not make it back to the southernmost states and Mexico disappear from the genetic lineage. In Florida, all stages from egg to adult can be found year-round.

The ova are green with vertical ridges like tiny monarch eggs, but the structure is very different under magnification. The caterpillars have spiny tubercles and are black with red, white, and blue markings. The colors are not as bright as many online photos suggest, but they are handsome creatures. The caterpillars feed on common plants like snapdragons and adventive plantains (*Plantago* spp.) as well as some native plants I have never seen or heard of before, like Canada toadflax and turkey tangle frogfruit. Caterpillars do not grow very large (<1.5″ or 3.8 cm) and require a limited amount of food. One small foodplant is enough to feed two or three to maturity. After hatching, they require about ten days of feeding and molting through five instars to reach the molt to the chrysalis.

The chrysalis is short and stumpy and is usually brown with white patches. The butterfly emerges after six days at room temperature. The liquid waste (or meconium) from a freshly emerged butterfly is red in color. This is usually released on the sides or bottom of the cage, but is also ejected as a defensive response to handling. As with most species, the meconium and the liquid waste produced after feeding appears the same, but can be a different color based on the food provided. The adults were offered red fruit jellies and artificial nectar from a flower dish with a sponge. Signs of feeding were visible on the surface of the jellies by the second day (dozens of tiny swipe marks). A week after emergence they were observed feeding from the flower dish.

Junonia coenia
adult, wings closed

Junonia orithya
The blue pansy, an Old
World species
(CC Michael MK Khor)

Junonia coenia
3rd and 4th instars

Junonia coenia
3rd instar ready to molt

Junonia coenia
5th instar caterpillars

Junonia coenia
chrysalis

Buckeye butterflies do not live a long time. My adults lived 37 to 50 days. As with the following migratory species, the short adult life seems to beg the question of how the expansive summer range and return impact the primary southern population.

PAINTED LADY BUTTERFLY
Vanessa cardui (Linnaeus, 1758)
Adults of this handsome, medium-size butterfly range from 2-2.5 ″ (5-6.4 cm) in wingspan. The painted lady is a member of the family Nymphalidae, which includes various angelwings, mourning cloaks, buckeyes, monarchs, and viceroys. It is a cosmopolitan species and is often said to be the most widespread butterfly species in the world. This species is likely the most widespread caterpillar found in classrooms and kept in homes. Painted lady butterfly kits are always available through biological supply houses, but are also seen for sale at garden shops, science stores, educator supply shops, local grocery stores, and even at close-out stores. The kit includes a net cage for the adults and a card to send away for caterpillars. Half a dozen or so caterpillars arrive in a cup with enough artificial diet to finish their life cycle.

Canadian thistle is an invasive plant found in flower beds and along roadsides across North America. Most readers may not know the name but are familiar with stepping on this painful weed when walking through a lawn barefoot. The plant is painful to pull from the ground without thick plastic gloves because the tiny needles break off under the skin and have to be dug out later with a sharp pin. Even when more than a foot of the root is pulled, it grows back. It even grows back after being treated with all but the strongest

Vanessa cardui
adults feeding

Vanessa cardui
(CC Renee Grayson)

herbicides. It is a joy to know something eats this terrible pest plant. The *V. cardui* in my photos were fed strictly on Canadian thistle. The caterpillars will also feed on other thistles, sunflower, and hollyhock.

Ova are tiny little blue specks about the size of the period at the end of this sentence. It is important to keep the leaves fresh or the eggs quickly dry out and die. They change from blue to black as the caterpillars develop inside. Hatching occurs just 48 hours after egg are laid.

Like the eggs, young caterpillars prior to fourth instar can die rapidly when the leaves dry out. Foodplant cuttings should be kept in a dram vial or other plastic container filled with water. A hole is poked in the lid of the vial and the stem wrapped with small pieces of paper towel to prevent curious little caterpillars from drowning. Cut thistle only lasts about four days before it starts to turn yellow and dry out. A fresh cutting should be placed against the old piece when it first begins to look a little yellow or withered at the tips. The caterpillars will crawl over to the fresh leaves. The larger caterpillars are most likely to find the fresh leaves without help. 1st and 2nd instar larvae often need to be carefully placed on fresh foodplant.

There are four molts, five instars. Black, 1.5 mm, 1st instar caterpillars begin to chew small holes in the undersides of the leaves. The top layer is left untouched so it looks like the leaves are being converted into wax paper (they only eat the bottom layer until the last instar). The very hungry caterpillars eat constantly and molt every 48 hours. 2nd and 3rd instar caterpillars spin silk webbing around the feeding area and will drop and hang from a silk thread when disturbed. 4th and 5th instar caterpillars drop to the ground and curl up to become a spiny ball. 5th instar is obtained in approximately eight days. 5th instar caterpillars are voracious feeders and food demands can be

Vanessa cardui
adults mating

Vanessa cardui
ova and early instars

Vanessa cardui
female laying eggs

Vanessa cardui
developing caterpillars

Vanessa cardui
caterpillars

Transparent leaves made
by feeding caterpillar

Vanessa cardui
chrysalis

difficult to keep up with. After three days the caterpillar climbs to the roof of the cage, or under the base of a foodplant leaf, and spins a silk pad. The back legs are hooked into the pad and then the caterpillar hangs head down until the molt into a chrysalis.

The chrysalis is tan with metallic gold spots. It wriggles energetically when disturbed but cannot go anywhere. This stage requires almost as much time as all the previous stages combined, about eight days. A few days before eclosion, the coloration of the adult butterfly begins to be visible through the transparent shell of the chrysalis.

Adults are easily fed in captivity in a small space (a large flight cage is unnecessary). The pop-up screen cage included in the kit is just nine inches (23 cm) in diameter and a little taller. The netting material is more important than the size. A cloth screen cage is best because the wings begin to look tattered after a week or two in a metal screen cage or aquarium. Butterflies readily drink from fresh orange slices but will also drink fluids such as sugar or honey water. After about ten days the adults mate and lay eggs. The adult stage is the longest part of the life cycle. Nevertheless, even when kept and fed well, these butterflies live just 21-35 days. In nature the adults migrate in huge numbers to Mexico each fall and repopulate most of North America every year (Glassberg 2011).

Black Swallowtail Butterfly
Papilio polyxenes Fabricius, 1775
This spectacularly beautiful butterfly is one of the more common swallowtails found around yards and fields. Caterpillars are also brightly colored but tend to be much more difficult to find than adults. Black swallowtails are big

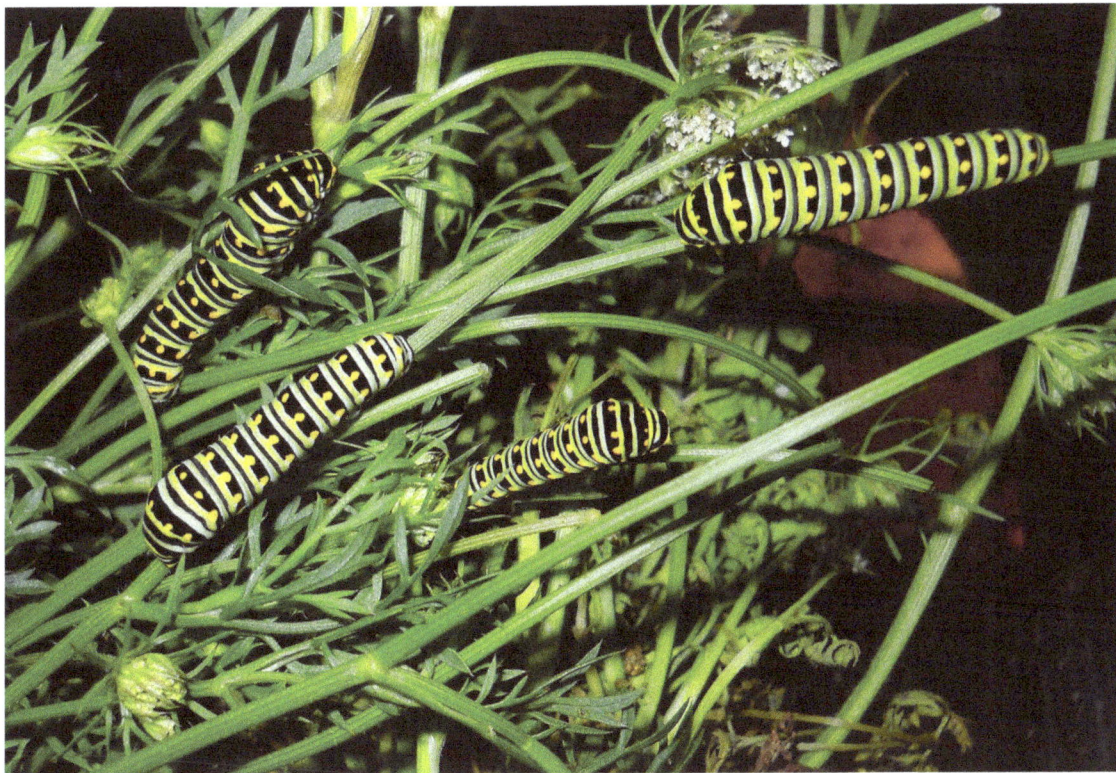

Papilio polyxenes 5th instar caterpillars on Queen Anne's lace

Small *Papilio polyxenes*
caterpillar looks like
a bird dropping

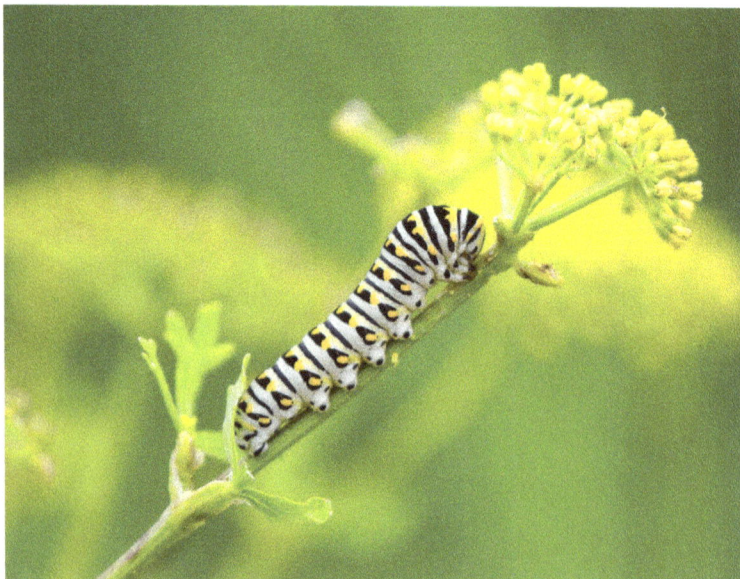

Papilio polyxenes
caterpillar
(CC Melissa McMasters)

butterflies, but are not as big as many other *Papilio* species. The maximum size is less than four and a half inches (<11.4 cm) from wingtip to wingtip. Oddly, the larval food of this North American native now seems to consist almost entirely of introduced European plants including fennel, parsnip, Queen Anne's lace (*Daucus carota*), and dill.

In summer 2012 I found five, tiny, 4 mm-long caterpillars on some dill seedlings growing next to a rock path in the yard. The combined seedlings could not have supported a single caterpillar to maturity. Five or ten years earlier I ran into caterpillars of this species that were also on a small group of very tiny dill plants, but they were eaten before I came back to check on them. This time I collected the larvae and dug up the small dill plants. There were enough dill leaves to get the caterpillars through one molt. I avoided buying dill plants at the local nursery in fear of pesticide residuals, so I switched to Queen Anne's Lace from the field behind my house. The plant is an extremely common field weed in my area, but I was not sure they would accept it because I have not encountered swallowtail caterpillars on *D. carota*. They all grew well, molted a few more times to 5th instar, and then to chrysalises.

When small, the caterpillars look like tiny bird droppings (giant swallowtail, *Papilio cresphontes*, larvae look like bird droppings for all instars). They transformed in color to become vivid green with black stripes in later instars. Large caterpillars are very beautiful, even more so in the defensive display. When disturbed they extrude the telltale swallowtail eversible defensive organ (the orange, odiferous, forked osmeterium) from just behind the head, on the prothorax. The osmeterium is easy to smell and observe momentarily, but getting the caterpillars to display it for more than a fraction of a second is

Papilio polyxenes
fresh chrysalis

Papilio polyxenes
chrysalis after six days

Papilio polyxenes freshly eclosed

difficult. I tried dozens of times to get a photo, but the osmeterium always retracted before the camera could focus.

Four chrysalises were formed on green stems and the fifth on a stained, brown piece of lumber. The surface color did not dictate the chrysalis color and they all started out bright green with a broken yellow stripe down the back. After a few weeks they turned dark brown but retained the yellow stripe.

The adults emerged after a month and were offered various foods. They refused sugar water from different containers and fruit jellies of various colors. They did not respond well to manual feeding. They began to fade after five days and were released into the garden where they had been found as caterpillars.

CABBAGE WHITE BUTTERFLY
Pieris rapae (Linnaeus, 1758)
This butterfly is also commonly known as the small white, though there are quite a few *Peiris* known as cabbage butterflies and they are mostly white. The other name for this butterfly, small white, refers to its smaller size compared to the large white *Pieris brassicae*, which is common in Europe. Members of this genus lay their eggs on cabbage and plants from the mustard family because the plant leaves contain sapponins, chemicals that are poisonous to many insects. Chemicals in these plants called glucosinolates are the cue for egg laying, and in their absence *Pieris rapae* females will lay no eggs.

The cabbage white is the most common species to fly around my yard; more than skippers and coppers combined. The adults are pretty and very easy to see at plus or minus 2 " (~50 mm). I could catch one almost any day

Pieris rapae male (top) and female (bottom)

between late spring and late summer. Still, I had not considered capturing adults to collect eggs and raise larvae. In July of 2012 I noticed a weed in a flower bed against the house with a dozen, green, inch-long caterpillars feeding on it. I normally encounter hundreds of small green "caterpillars" on the plants around the house ever year, but they are almost always sawflies. I knew these would become butterflies so I decided to keep an eye on them. However, as usually happens with caterpillars left to the loving care of nature, a few days later they were all gone. Usually missing caterpillars are the direct result of predators, but in this case they finished off the tiny plant's leaves and may have simply wandered off.

I noticed the same type of weeds in a patch of grass near my work at the end of September. I inspected the plants and sure enough there were more caterpillars and even a few of the very tiny ova. Since there was very little leaf material available (the weeds were small) I bought a head of cabbage from the grocery store. They switched to this food with no hesitation.

The small, bright green caterpillars are said to hide under leaves, but the ones I collected outdoors primarily sat on the top side of the stem in plain view. Nevertheless, they are invisible if you are not looking for them. The plants they were found on are common weeds introduced from Europe known as wintercress (*Barbarea vulgaris*). The caterpillars grew quickly and within a few days the largest had become chrysalises. The chrysalises are brighter green than the caterpillars and take about two weeks to develop into butterflies.

Male butterflies have just one black spot near the middle of the forewings, whereas females have two, one over the other. I offered the adults beetle jellies flavored banana and brown sugar, but neither convinced the butterflies

Pieris rapae eggs

Pieris rapae caterpillar

Pieris rapae
chrysalis

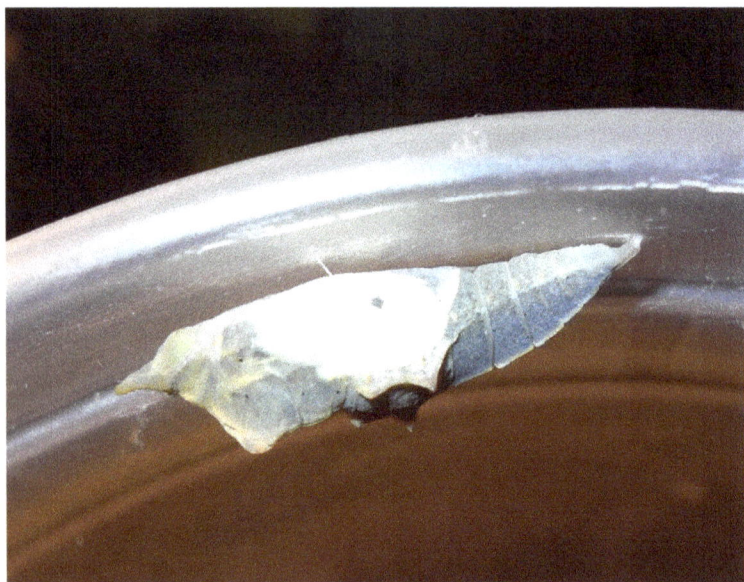

Pieris rapae chrysalis
ready to eclose

to unroll a single proboscis when placed on the food. Honey water was consumed during manual feedings. Honey water cups were left in the cage, but I am not certain they were able to locate it on their own.

Adults began to lay eggs just two days after emergence from the chrysalis. Eggs are tiny, yellow, spindle-shaped, and glued singly to the leaf, like a tiny thorn. A few eggs are laid at a time, but in the confines of captivity numerous eggs were laid over time on a single leaf. Unfortunately the hundred or so eggs laid over the next two weeks appear to have all been infertile. Each adult only lived around three weeks.

The neatest thing I learned from keeping this species is that leaves torn off a cultivated cabbage head and placed in water will grow a ton of roots and even leaf growth buds.

Pandorus sphinx moth, *Eumorpha pandorus*

CLOSING

The maintenance of Lepidoptera in artificial habitats is an educational experience that can last a lifetime. Observing the transformation of eggs that come in a huge array of shapes, sizes, and colors, into equally variable caterpillars, can become an endless pursuit for the lepidopterist. Many caterpillars transform again into very different looking creatures as they molt into each instar. The chrysalis and cocoon can also be fantastically colored or constructed. All this complexity and beauty and the animal still has not reached the pinnacle of development familiar to the masses. Butterflies and moths are well-known for their beauty.

A book 1,000 times the size of this text would not be able to fully detail everything known on the biology and husbandry for this group of insects. *Lepidopteran Zoology* showcases the joy of observing life cycles of the amazing Lepidoptera through a handful of representative species. Hopefully I have provided an adequate framework to aid the beginning lepidopterist's zoological adventures and shared my enthusiasm and love for these creatures. The best way to fully appreciate butterflies and moths is not through amassing a collection of beautiful, dried adult specimens, but through observations of their growth and development. If this book inspires just one person to work with butterflies and moths, then this effort was a spectacular misuse of time. Nevertheless, writing about, studying, documenting, and recalling experiences has been incredibly fun.

Io moth, *Automeris io*
(CC Andy Reago and Chrissy McClarren)

GLOSSARY

ABDOMEN: The terminal section of the insect body, contains the gut, anus, and genital organs.

CLASPERS: Male appendages near the end of the abdomen that hold the female during mating.

CLYPEUS: Wide plate on an insect head that holds the mandibles and labrum.

DESICCATION: Drying out, often to the point of death.

DIAPAUSE: A period of inactivity or arrested development influenced by seasonal changes in temperature, light cycle, or humidity, such as overwintering.

ECLOSE: To emerge as an imago from the chrysalis or pupa.

EXOSKELETON: The external body covering of arthropods which serves as protection and muscular support. Mainly consisting of chitin and proteins.

FEMUR (*pl.* FEMORA): Section of the insect leg located between the trochanter and tibia; usually the largest segment of the leg.

FRONS: Triangular section of the head between the lobes and above the clypeus.

GREGARIOUS: Living in groups or a preference to be near others of one's own kind.

HEAD CAPSULE: Round, hard shell containing the mouthparts and eyes; common feature of lepidopteran, coleopteran, and hymenopteran larvae.

INSTAR: Developmental stage between each molt and prior to maturity.

LABRUM: Structure corresponding to a lip for insect mouthparts.

LOBES: Large, paired frontal areas of the caterpillar head above the frons.

MARGIN: A border or edge. In lepidopteran wings the margin is the edge of the side and bottom.

MECONIUM: Liquid waste of a newly emerged butterfly or moth.

MOLT: *verb.*, To shed the exoskeleton or skin. *noun.*, the act of shedding or replacing the outer shell or skin; the discarded skin/shell.

OCELLUS (*pl.* OCELLI): Light detection organs. Caterpillars usually have six ocelli near the bottom edge of each lobe.

OSMETERIUM: Eversible defensive organ that emits unpleasant odors; only found on swallowtail caterpillars.

OVIPOSITION: The laying or deposition of eggs.

OVUM (*pl.* OVA) Egg.

PARASITOID: An organism that feeds off a single animal and kills its host, as opposed to a parasite that does not usually kill its host.

PARTHENOGENESIS: Reproduction through the development of unfertilized eggs.

PROBOSCIS: Lepidopteran adult mouthparts used to drink fluids.

PROLEG: Leg-like appendages on the third through sixth abdominal segment of caterpillars.

SETA (*pl.* SETAE): Hollow growth from the epidermal layer of the body wall: hair-like setae are often referred to as "hair" but are principally different from the hair of mammals. Setae are principal to the appearance and defense of many caterpillars, but on adult Lepidoptera most are flattened and widened into scales.

SEXUALLY DIMORPHIC: Presence of noticeably different external gender characteristics including size and color.

STRIDULATE: To create noise by rubbing together body parts, usually specialized legs, mouthparts, or wings.

TENERAL: Soft, often pale colored, condition of the insect exoskeleton immediately following a molt.

TUBERCLE: Bumpy outgrowths of the exoskeleton. In caterpillars the location of these fleshy outgrowths can be described as caudal (top middle), dorsal (top sides), and lateral (sides above legs).

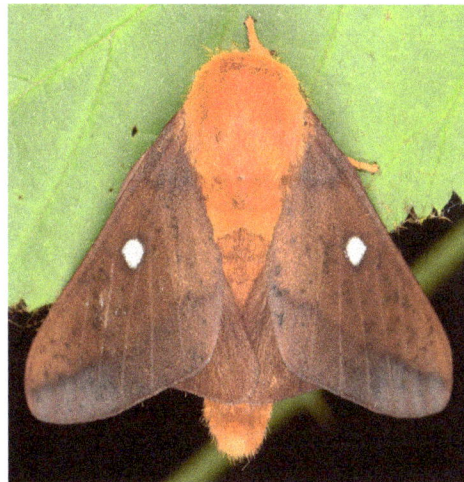

Anisota peigleri
(CC Christina Butler)

Luna moth, *Actias luna*, discovered by my wife on the front porch

BIBLIOGRAPHY

Arnett, Ross H. (1993) *American Insects A Handbook of the Insects of America North of Mexico*. Gainesville, Florida: Sandhill Crane Press.

Eliot, Ida, and Caroline C. Soule. (1921) *Caterpillars and Their Moths*. New York: The Century Co.

Gardiner, Brian O. C. (1982), Crotch (1956) and Copper (1942) *A Silkmoth Rearer's Handbook*. Middlesex: The Amateur Entomologist's Society.

Glassberg, Jeffrey. (2011) *Butterflies of North America*. New York: Sterling Publishing Co.

Johnson, Sylvia A. (1982) *Silkworms*. Minneapolis, MN: Lerner Publications Company.

McMonigle, Orin. (2003) Monarch of the butterflies. *Invertebrates-Magazine* 2(2): 14-17.

McMonigle, Orin. (2005) The cynthia moth *Samia cynthia*. *Invertebrates-Magazine* 5(1): 9-12.

McMonigle, Orin. (2007) Promethea: a moth to admire. *Invertebrates-Magazine* 6(2): 12-15.

McMonigle, Orin. (2007b) Featured invertebrate: tobacco hornworm sphinx moth *Manduca sexta*. *Invertebrates-Magazine* 6(4): 18.

McMonigle, Orin. (2009) Featured invertebrate: cecropia moth *Hyalophora cecropia*. *Invertebrates-Magazine* 8(4): 16-17.

McMonigle, Orin. (2010) Too many molts and the gargantuan tentless tent caterpillar. *Invertebrates-Magazine* 9(2): 4-7.

McMonigle, Orin. (2011) Featured invertebrate: silkworm *Bombyx mori*. *Invertebrates-Magazine* 10(4): 12-13.

McMonigle, Orin. (2011b) *Invertebrates for Exhibition: Insects, Arachnids, and Other Invertebrates Suitable for Display in Classrooms, Museums, and Insect Zoos*. Landisville, PA: Coachwhip Publications.

McMonigle, Orin. (2012) Summer of butterflies. *Invertebrates-Magazine* 11(2): 14-17.

McMonigle, Orin. (2012b) Featured invertebrate: captive life cycle of the milkweed tussock moth *Euchaetes egle* (Drury, 1773). *Invertebrates-Magazine* 12(1): 15-17.

McMonigle, Orin. (2018) Rearing the white-lined sphinx *Hyles lineata*. *Invertebrates-Magazine* 18(1): 6-9.

McMonigle, Orin. (2019) Life cycle of the common buckeye butterfly *Junonia coenia* Hübner, *Invertebrates-Magazine* 18(2): 7-9.

Merian, Maria Sibylle. (1679) *Der Raupen Wunderbare Verwandlung und Sonderbare Blumennahrung [The Caterpillars' Marvelous Transformation and Strange Floral Food]*. Vol. 1. Nuremberg: Johann Andreas Graff.

Mouffett, Thomas. (1599) *The Silkewormes and Their Flies*. London: Nicholas Ling.

Villiard, Paul. (1969) *Moths and How to Rear Them*. Funk & Wagnalls, New York.

Warrington, R. (1841) Art. XIV. Notice of the peculiar economy of certain larvae, in eating the egg-shell which previously contained them. *The Entomologist* 1(April): 96.

Watts, R. C. (1996) Hand-pairing swallowtails. *Bulletin of the Amateur Entomologists' Society*. 44(405): 78.

Pandora moth, *Eumorpha pandorus*, caterpillar
(CC Scott King)

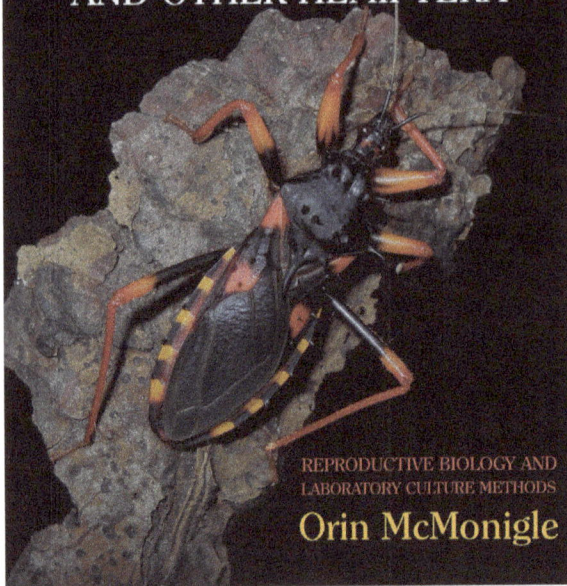

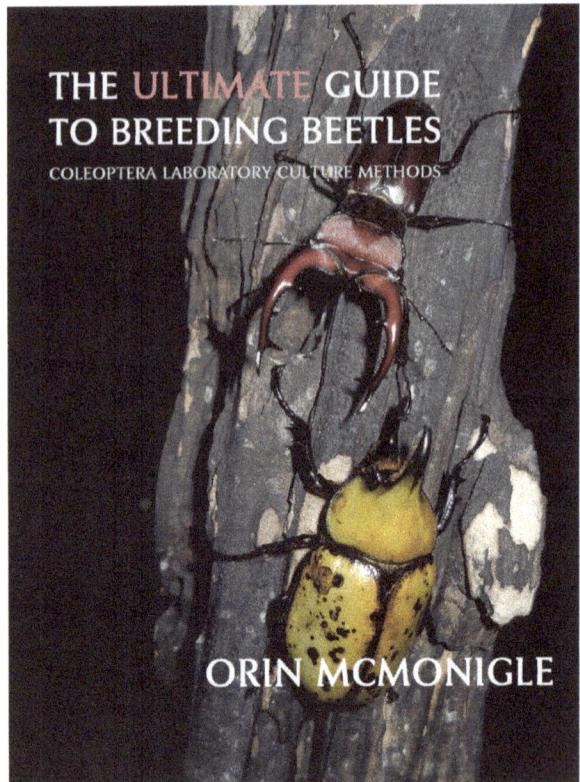

FOR THE LOVE OF
COCKROACHES

蜚蠊之愛

ORIN MCMONIGLE

**KEEPING THE
PRAYING MANTIS**

Orin A. McMonigle

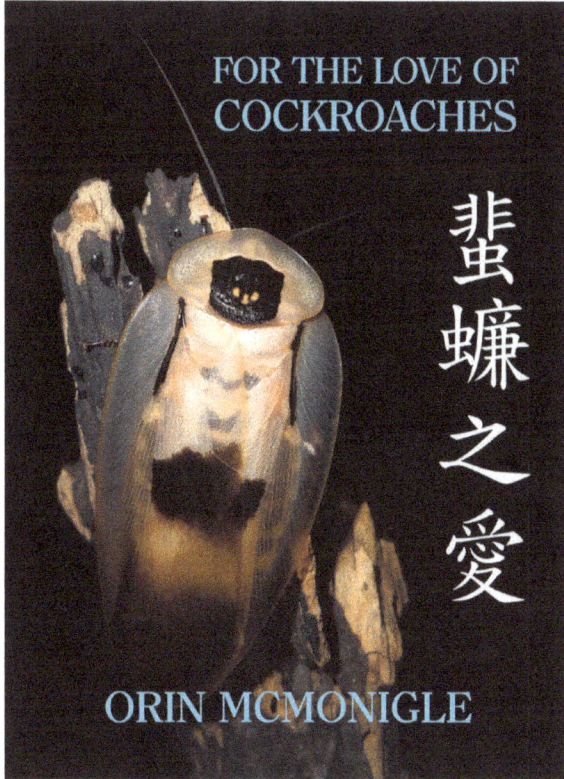

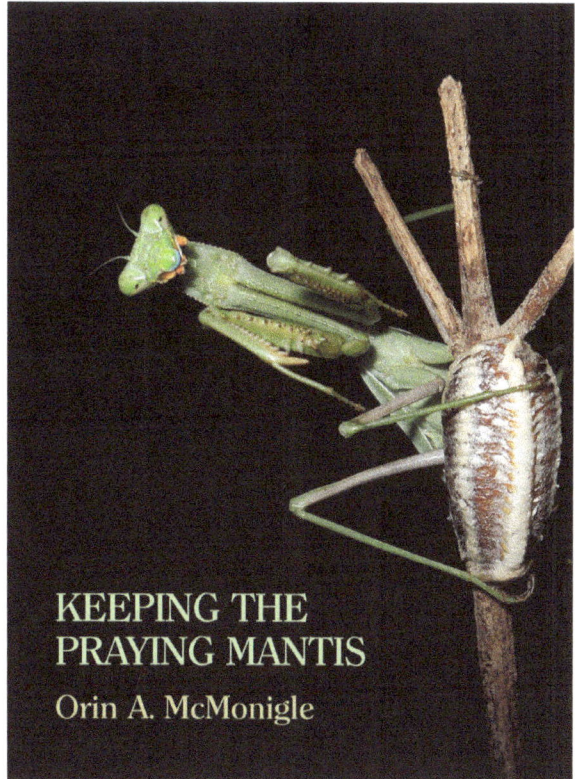

ALSO AVAILABLE FROM
COACHWHIP PUBLICATIONS

CoachwhipBooks.com

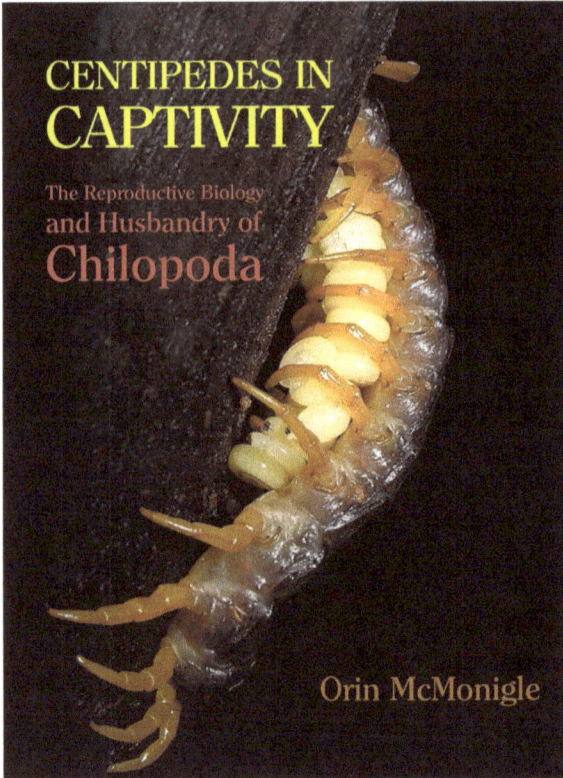

CENTIPEDES IN CAPTIVITY

The Reproductive Biology
and Husbandry of
Chilopoda

Orin McMonigle

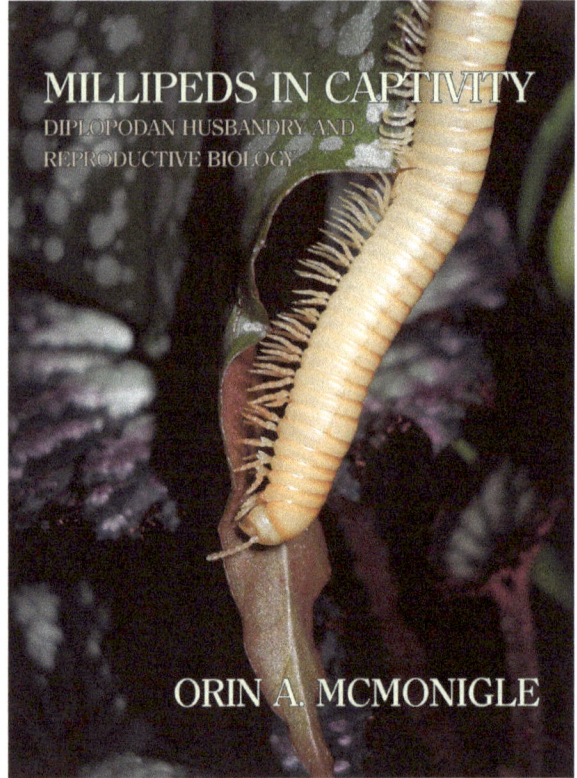

MILLIPEDS IN CAPTIVITY
DIPLOPODAN HUSBANDRY AND
REPRODUCTIVE BIOLOGY

ORIN A. MCMONIGLE

ALSO AVAILABLE FROM
COACHWHIP PUBLICATIONS

COACHWHIPBOOKS.COM

BREEDING THE WORLD'S
LARGEST LIVING ARACHNID
AMBLYPYGID BIOLOGY, NATURAL HISTORY, AND CAPTIVE HUSBANDRY
ORIN MCMONIGLE

FORGOTTEN
ORDER OF THE
VINEGAROONS
Orin McMonigle

WHIPSCORPION
BIOLOGY, HUSBANDRY,
AND NATURAL HISTORY

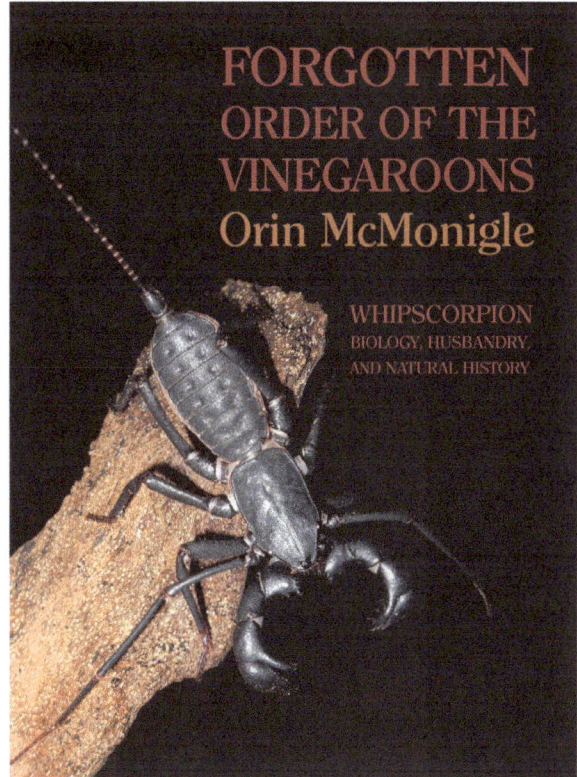

ALSO AVAILABLE FROM
COACHWHIP PUBLICATIONS

CoachwhipBooks.com

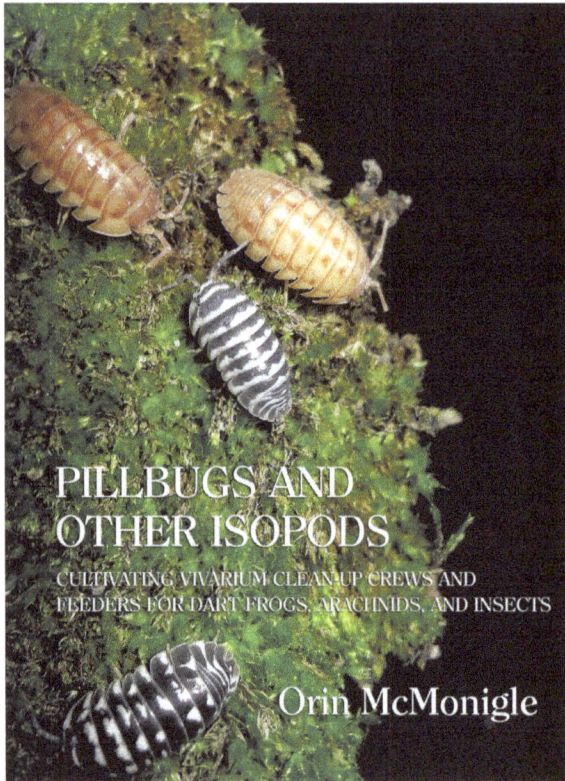

PILLBUGS AND OTHER ISOPODS

CULTIVATING VIVARIUM CLEAN-UP CREWS AND FEEDERS FOR DART FROGS, ARACHNIDS, AND INSECTS

Orin McMonigle

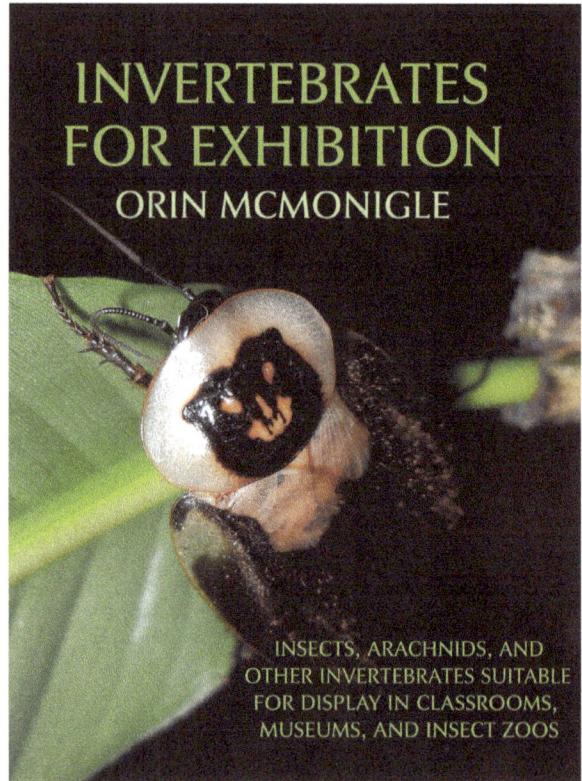

INVERTEBRATES FOR EXHIBITION

ORIN MCMONIGLE

INSECTS, ARACHNIDS, AND OTHER INVERTEBRATES SUITABLE FOR DISPLAY IN CLASSROOMS, MUSEUMS, AND INSECT ZOOS

ALSO AVAILABLE FROM
COACHWHIP PUBLICATIONS

COACHWHIPBOOKS.COM

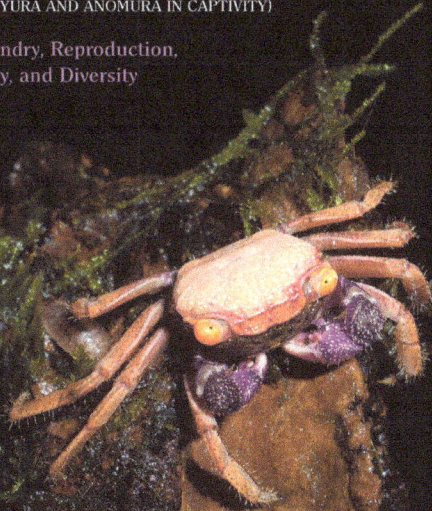

BREEDING THE VAMPIRE AND OTHER CRABS

(BRACHYURA AND ANOMURA IN CAPTIVITY)

Husbandry, Reproduction, Biology, and Diversity

ORIN MCMONIGLE

ALSO AVAILABLE FROM
COACHWHIP PUBLICATIONS

CoachwhipBooks.com

www.ingramcontent.com/pod-product-compliance
Lightning Source LLC
Chambersburg PA
CBHW040247290326
41929CB00054B/3450